Advances in Battery Manufacturing and Operating Status Analysis

IEEE Press
445 Hoes Lane
Piscataway, NJ 08854

Advances in Battery Manufacturing and Operating Status Analysis

Filtering and Artificial Intelligence Strategy

Ziyun Wang
School of Automation and Intelligent Science, Jiangnan University Wuxi, Jiangsu Province, P. R. China

Yan Wang
School of Automation and Intelligent Science, Jiangnan University Wuxi, Jiangsu Province, P. R. China

Zhicheng Ji
School of Automation and Intelligent Science, Jiangnan University Wuxi, Jiangsu Province, P. R. China

IEEE Press
WILEY

Published by John Wiley & Sons, Inc., Hoboken, New Jersey.

Library of Congress Cataloging-in-Publication Data has been applied for:

Print ISBN: 9781394437160
ePDF ISBN: 9781394437184
ePub ISBN: 9781394437177
oBook ISBN: 9781394437191

Cover Design: Wiley
Cover Image: xPACIFICA/Getty Images

To my beloved Haiying Feng and our children, your love is the light behind every page.

Contents

List of Figures

List of Tables

About the Authors

Ziyun Wang received the PhD degree in control science and engineering from Jiangnan University, Wuxi, China, in 2015. He is a Professor at Jiangnan University and currently the Deputy Director of the Engineering Research Center of Internet of Things Technology Applications (Ministry of Education), and Deputy Director of the School of Automation and Intelligent Science, Jiangnan University. His current research interests include filter design, state estimation, and their applications in battery manufacturing and operating status analysis.

Yan Wang received the PhD degree in control theory and control engineering from Nanjing University of Science and Technology, Nanjing, China, in 2006. From 2013 to 2014, she was a Visiting Researcher with the School of Electrical and Computer Engineering, Louisiana State University, Baton Rouge, LA, USA. She is currently a Professor at Jiangnan University, the Head of the Provincial Excellent Science and Technology Innovation Team, and a recipient of the Yangtse River Scholar of the Ministry of Education of China. Her research interests include complex industrial process modeling and intelligent manufacturing.

Zhicheng Ji received the PhD degree in power electronics and drives from China University of Mining and Technology, China, in 2004. From 2003 to 2004, he was a Visiting Researcher with the University of Toronto, Canada. He is currently a Professor at Jiangnan University, a National Teaching Master, and a fellow of the Asian Simulation Alliance. He has been awarded the First Prize of Science and Technology Progress Award by the Ministry of Education of China in 2011 and 2016, and the First Prize of National Teaching Achievement Award in 2018. His research interests include artificial intelligence, intelligent manufacturing of new energy batteries, and control techniques for the Internet of Things.

Preface

As the global energy transition accelerates, battery technology has become central to achieving clean energy goals. The performance, safety, and longevity of batteries are critical for the widespread adoption of new energy technologies. However, the increasing complexity of battery systems presents significant challenges in managing and optimizing their manufacturing, operation, and health status. Traditional methods, while foundational, are limited when dealing with the complexity, noise, and uncertainty inherent in modern battery systems. This book addresses this gap by integrating advanced filtering techniques with artificial intelligence (AI) strategies, driving the smart development of battery technology. We provide a unified framework that combines data-driven algorithms with conventional battery management and state estimation technologies, offering new solutions for optimizing the entire battery lifecycle.

The book covers key areas of battery manufacturing and management. We begin with AI applications in battery grouping and manufacturing capacity forecasting, introducing strategies based on genetic algorithms and Q-learning to improve production efficiency and consistency. We then explore AI-driven optimization of state of charge (SOC) estimation, even in the presence of significant noise, ensuring greater accuracy and stability. As the book progresses, we delve into advanced filtering methods, including zonotopic Kalman filtering (ZKF) and particle swarm optimization (PSO), which tackle the complexities of battery systems and improve the precision of battery status analysis. The integration of AI and filtering techniques for predicting remaining useful life (RUL) is also emphasized, demonstrating how hybrid models can accurately predict battery health and failure modes, enabling predictive maintenance. These methods provide powerful support for efficient and reliable battery management, even under challenging operational conditions.

This book is primarily aimed at senior undergraduate students, graduate students, and researchers in the fields of control theory, automation, intelligent manufacturing, and related disciplines. Additionally, for engineers and technicians involved in battery management system design and optimization, the theoretical analysis and practical case studies presented in this book will also serve as valuable references.

It is hoped that this book will provide readers with new perspectives and ideas in the research and practice of battery technology, state estimation, and health management.

Ziyun Wang
Jiangnan University
Yan Wang
Jiangnan University
Zhicheng Ji
Jiangnan University

Acknowledgments

The authors wish to express sincere gratitude to all who have supported and encouraged us during the writing of this monograph. We are deeply thankful to colleagues and research group members for their stimulating discussions, constructive feedback, and continuous support. Special thanks go to the anonymous reviewers and the editorial team, whose insightful comments and professional guidance greatly improved the quality of this work. We also acknowledge the administrative and technical staff at our institution for their indispensable assistance. This monograph is supported by the Natural Science Foundation of China with Grant No. 62473174 and the Natural Science Foundation of Jiangsu Province with Grant No. BK20221533.

Symbol Description

$\mathbb{R}$	Set of real numbers
$\mathbb{R}^n$	Set of n-dimensional real vector
$\mathbf{B}^n$	Unitary box in $\mathbb{R}^n$
A^{T}	Transpose of matrix A
$\hat{A}$	Estimated value of matrix A
A^{-1}	Inverse of matrix A
$\det(A)$	Determinant of matrix A
$\mathbf{tr}(A)$	Trace of matrix A
$A > 0$	General notation for strictly positive definite matrix A
$A \geq 0$	General notation for positive definite matrix A
$A < 0$	General notation for strictly negative definite matrix A
$A \leq 0$	General notation for negative definite matrix A
$\boldsymbol{I}_n$	Identity matrix in $\mathbb{R}^{n\times n}$
$\mathbf{0}_n$	Zeros matrix in $\mathbb{R}^{n\times n}$
$\mathrm{diag}(a_1, \dots, a_n)$	Diagonal matrix of dimension n
$\lvert \cdot \rvert$	Absolute value
$\lVert \cdot \rVert_\infty$	Infinity norm
$\lVert \cdot \rVert_P$	P-norm
$\lVert \cdot \rVert_F$	Frobenius norm
s.t.	Subject to
$\in$	Belongs to
$\notin$	Not belongs to
$\subset$	Subset
$\supset$	Contained in
$\cap$	Intersection
$\oplus$	Minkowski sum
$\odot$	Linear mapping
$\mathbf{conv}(\cdot)$	Convex hull
max	Maximum value
min	Minimum value
$\emptyset$	Empty set

1

Introduction

1.1 Research Background and Significance

The global transition toward carbon neutrality has spurred rapid advancements in industries such as new energy vehicles, large-scale energy storage systems, and smart grids, all of which rely heavily on high-performance, safe, and long-lasting batteries [1]. Lithium-ion batteries (LIBs), renowned for their high energy density, long cycle life, and efficient rechargeability, have become indispensable for these applications [2]. As the demand for energy storage and electric vehicles (EVs) continues to surge, driven by both environmental concerns and technological progress, the battery industry faces unprecedented pressure to meet the growing need for reliable and efficient power sources [3].

However, despite the significant progress in battery technologies, several persistent challenges hinder the development and optimization of battery systems [4]. On the manufacturing side, the industry faces a fundamental mismatch between the increasing demand for batteries and the inconsistency in manufacturing processes. Battery production processes often struggle with issues such as low efficiency, high production costs, and challenges in predicting manufacturing capacity under nonstationary conditions, which can adversely affect both product quality and operational performance [5]. Achieving high manufacturing consistency and reliability is now one of the most pressing concerns in the battery manufacturing sector. Additionally, the increasing complexity of battery designs, such as those for EVs and large-scale energy storage, demands novel approaches to manufacturing, such as more precise grouping methods and optimized production forecasting [6].

From an operational perspective, ensuring the reliable performance of batteries, accurately predicting their remaining useful life (RUL), and estimating their state of charge (SOC) are critical challenges that have not been fully addressed by traditional methods [7]. These issues are further complicated by environmental variability, aging effects, and nonlinear degradation processes, which introduce significant uncertainty. Traditional state estimation methods, such as those based on Kalman filtering, typically rely on assumptions of Gaussian noise and known system models, which are often incompatible with real-world applications where noise is unknown and time-varying. This mismatch between theoretical assumptions and practical realities underscores the need for more robust and adaptive estimation techniques capable of handling complex and uncertain battery systems.

To address these challenges, the integration of advanced filtering techniques and artificial intelligence (AI) strategies offers promising solutions [8]. Filtering methods, such as zonotopic filtering and particle filters (PFs), provide a solid framework for dealing with uncertainties and noise in battery data. These methods enable more accurate state estimation, particularly in scenarios where the underlying system exhibits nonlinearities and noise cannot be assumed to follow a Gaussian distribution. PFs, in particular, approximate the posterior probability distribution of the system state by using a set of random samples ("particles"), making them well-suited for complex and uncertain environments, such as those encountered in battery systems [9]. In addition to filtering techniques, AI strategies, including deep learning (DL) algorithms like long short-term memory (LSTM) networks and convolutional neural networks (CNNs), have been increasingly applied to battery systems for tasks such as forecasting manufacturing capacity, predicting RUL, and improving SOC estimation [4, 10]. These AI-driven approaches are adept at capturing nonlinear patterns and long-range dependencies in data, making them particularly effective for battery state estimation and performance prediction [11]. By combining the strengths of advanced filtering and AI techniques, this book aims to provide novel solutions to the key challenges faced by the battery industry, driving the transition from traditional manufacturing methods to "smart manufacturing" processes that are more efficient, reliable, and scalable.

The significance of this work lies not only in advancing the theoretical understanding of these techniques but also in providing practical insights that can be applied to real-world battery systems. Through the integration of cutting-edge filtering methods and AI strategies, this book offers a comprehensive approach to enhancing battery performance and optimizing manufacturing processes, which is crucial for meeting the increasing demand for high-quality, reliable batteries in the face of growing global energy needs. Ultimately, this research contributes to the long-term goal of creating smarter, more efficient battery systems that can support a sustainable future.

1.2 Battery Manufacturing

Battery manufacturing is a highly intricate process that encompasses the production of various components, including electrodes, separators, and electrolytes, which are subsequently assembled into functioning energy storage systems [5].

The complexity of this process is further heightened by the stringent requirements for uniformity, high performance, and safety, particularly in applications such as EVs and large-scale energy storage systems. As the demand for advanced batteries continues to rise, manufacturers face the challenge of maintaining high production standards while simultaneously increasing production capacity and reducing costs.

A primary concern in battery production is ensuring the uniformity of the cells. Variations in materials and manufacturing processes, such as discrepancies in electrode coating thickness, separator porosity, or electrolyte composition, can lead to significant differences in battery performance, cycle life, and safety. Such variations can also compromise the long-term reliability of the batteries, making precise quality control essential throughout the entire manufacturing process. Traditional methods of ensuring uniformity often rely on manual inspection and empirical techniques, which are not only time-consuming but can also be prone to error. As a result, there is an increasing interest in automating the manufacturing process and applying more advanced techniques such as machine learning (ML) and AI to achieve greater precision and consistency in production.

Another major challenge in battery manufacturing is predicting and optimizing production capacity, particularly when the manufacturing environment is subject to variability. The production process is often nonstationary, meaning that the data generated during production, such as temperature, pressure, and voltage, can fluctuate over time due to factors like equipment aging, material degradation, and environmental conditions. This variability complicates the task of forecasting production efficiency and capacity. Traditional methods of capacity prediction are ill-equipped to handle such fluctuations, leading to inefficiencies in production scheduling and resource allocation. To address these issues, recent studies have explored the use of data-driven models, such as those incorporating trend filtering segmentation under fuzzy information granulation, which can adapt to changes in the production environment and provide more accurate predictions [6]. These models allow for dynamic adjustments in response to variations in the production process, thus enabling more reliable capacity forecasting.

Furthermore, the grouping of battery cells based on their characteristics, such as voltage, capacity, and internal resistance, is a critical step in ensuring the optimal performance of battery packs. Poorly matched cells can result in imbalanced charge/discharge cycles, reduced capacity, and increased risk of thermal runaway or other safety hazards. While traditional grouping methods typically involve manual sorting and empirical rules, these approaches are time-consuming and often ineffective. Recently, AI-based algorithms, such as genetic algorithms and active equalization methods based on clustering analysis, have been proposed to automate the cell grouping process [12]. These methods improve the accuracy and efficiency of grouping by using optimization techniques to identify the best match between cells, thus enhancing the overall performance and longevity of the battery pack.

In addition to these technical challenges, battery manufacturers also face increasing pressure to meet both environmental and economic goals. As the demand for EVs and renewable energy systems grows, manufacturers must adopt cost-effective production methods that do not compromise quality or performance [3]. This is especially critical in a highly competitive global battery market, where cost reduction and environmental sustainability are key priorities. The integration of advanced technologies such as AI-driven optimization, robotics, and automated quality control systems is enabling manufacturers to improve efficiency, reduce energy consumption, and minimize waste generation during production. These innovations contribute to the environmental sustainability of the industry while simultaneously enhancing the economic feasibility of battery manufacturing.

In conclusion, battery manufacturing remains a complex and evolving field, with significant challenges related to production consistency, capacity forecasting, and performance optimization. The application of advanced filtering techniques, AI-driven models, and automation presents promising solutions to these challenges, enabling manufacturers to meet the growing demand for high-performance batteries. The future of battery manufacturing will likely see the continued adoption of smart manufacturing techniques that leverage real-time data, AI, and automation to optimize production efficiency and quality.

1.3 Battery Operating Status Analysis

The analysis of battery operating status plays a crucial role in ensuring the safe and efficient operation of battery-powered systems, particularly in applications such as EVs, renewable energy storage, and portable electronic devices. The primary goal of battery status analysis is to estimate key parameters, such as the SOC, state-of-health (SOH), and RUL, which are essential for monitoring the battery's performance and predicting potential failures [2]. These parameters are influenced by various factors, including operational conditions, aging effects, and environmental influences, making their accurate estimation a challenging task.

One of the most critical parameters in battery operating status analysis is the SOC, which indicates the remaining energy in the battery and is essential for efficient operation. Accurate SOC estimation ensures that the battery operates within safe limits, preventing overcharging or deep discharging, both of which can significantly reduce the battery's lifespan. However, SOC estimation is complicated by the nonlinear behavior of batteries, which are influenced by factors such as temperature, charge/discharge rates, and internal resistance. Traditional SOC estimation techniques, including the open-circuit voltage (OCV) method and coulomb counting, are often insufficient in dynamic environments where the battery's behavior changes due to aging or external conditions.

Recent advances in battery state estimation have seen significant improvements through robust filtering strategies. While Kalman filters (KFs) provide recursive state estimation from noisy measurements, their assumption of linear state-measurement relationships often proves inadequate for strongly nonlinear battery dynamics. More flexible approaches like particle filtering approximate the posterior probability distribution using random samples ("particles") [9]. Beyond probabilistic methods, set-membership filtering employing zonotopes—convex polytopes efficiently representing multidimensional uncertainty sets—has demonstrated particular strength in estimating SOC under unknown but bounded (UBB) and time-varying noise conditions. Zonotopic filtering models state uncertainty through a center vector and generator matrix, using linear mappings and Minkowski sums for prediction and update steps without requiring specific noise distributions. This framework not only provides point estimates but also computes guaranteed state bounds, offering a computationally efficient and robust solution for battery management systems (BMSs) operating under significant uncertainty.

Alongside SOC, SOH is another essential parameter that reflects the overall condition of a battery. SOH is typically determined by factors such as internal resistance, capacity fade, and voltage response, which change as the battery ages. Accurate SOH estimation is vital for understanding the degradation of a battery over time, as it helps predict when the battery will no longer perform at optimal levels. Conventional methods of SOH estimation often rely on empirical models or simplified battery aging models, which may fail to capture the complex degradation processes occurring in real-world applications.

To overcome these limitations, more advanced techniques have been proposed, combining both model-based and data-driven approaches. DL methods have demonstrated the ability to estimate lithium-ion battery SOH without additional degradation experiments [10]. Hybrid DL methods have also been developed for the estimation of the SOH of LIBs, showing improved accuracy and robustness [4]. Battery degradation prognosis and health management based on grey iterative modeling provides another innovative approach to SOH estimation [13]. These hybrid methods, which combine model-based and data-driven approaches, have proven effective in improving SOH estimation by leveraging the strengths of both techniques.

Finally, predicting the RUL of a battery is another critical aspect of battery operating status analysis. RUL prediction is important for determining when a battery will no longer be able to perform its intended function, thus enabling timely maintenance or replacement. RUL prediction is particularly challenging due to the complex and nonlinear nature of battery aging, which is influenced by numerous factors such as charge/discharge cycles, temperature, and operating stresses. Traditional RUL prediction models often fall short of capturing the time-varying nature of the battery degradation process.

Recent advancements have significantly improved the accuracy of RUL prediction. Techniques such as particle filtering and neural networks, especially LSTM networks, have demonstrated exceptional performance in predicting RUL. Survival analysis combined with ML has emerged as a powerful framework for predicting Li-ion battery RUL [11]. These models are capable of learning complex temporal dependencies and nonlinear patterns in the battery data, which is essential for predicting the RUL accurately. By training these models on large datasets, they can effectively predict the remaining lifespan of a battery based on historical data and operational conditions.

In conclusion, battery operating status analysis is fundamental for ensuring the efficient and safe operation of battery systems across a wide range of applications. The accurate estimation of SOC, SOH, and RUL is crucial for optimizing battery performance, enhancing safety, and extending the battery's lifespan. The integration of advanced methods, such as particle filtering, ML algorithms, and hybrid techniques, provides robust solutions to the challenges of battery status analysis. These methods not only improve the accuracy of status estimation but also enable more effective monitoring and management of battery systems in real-world environments.

1.4 Principle of Filter Design Methods

Filter design methods are essential tools for addressing the challenges of state estimation, particularly in systems characterized by uncertainty and noise. In battery systems, these methods are crucial for accurately estimating parameters such as SOC, SOH, and RUL, which are often affected by various disturbances and uncertainties, including measurement noise, model inaccuracies, and environmental factors. A filter is an algorithm that processes a sequence of noisy observations to estimate the state of a dynamic system. The objective of filter design is to reduce the uncertainty in the system's state estimation by using mathematical models that describe the system's dynamics and measurement process.

The design of effective filters for battery systems involves several fundamental principles that stem from various fields, including linear and nonlinear systems, estimation theory, and statistical inference. These principles are broadly categorized into linear filters, nonlinear filters, and robust filters, each offering different advantages and applicable scenarios.

1.4.1 Linear Filters

Linear filters are among the most commonly used and simplest types of filters in state estimation. They assume that the system and measurement models are linear, and that the noise affecting the system is Gaussian. The KF is the most widely used linear filter and is optimal in the sense that it minimizes the mean square error of the state estimate, given the assumptions of Gaussian noise and linearity. In the KF, the system's state is updated iteratively using both the previous state estimate and the new measurements, and it provides an estimate of the state along with a covariance matrix that quantifies the uncertainty in the estimate.

The extended Kalman filter (EKF) is an extension of the KF that deals with systems exhibiting nonlinear behavior. It linearizes the system's nonlinear dynamics around the current estimate, enabling the use of KF principles for nonlinear systems. However, EKF has limitations in handling highly nonlinear systems, as it may introduce errors in the linearization process.

1.4.2 Nonlinear Filters

While linear filters work well for systems with linear dynamics and Gaussian noise, they are often inadequate for handling nonlinear systems, which are common in battery modeling due to the complex electrochemical processes involved. PF, also known as sequential Monte Carlo (SMC) methods, represent a class of filters that do not require linearity assumptions. PFs represent the system's state with a set of random samples, or "particles," each of which has a weight based on its likelihood. The filter updates the particles over time according to the system's dynamics and measurement model, and the state estimate is obtained by calculating the weighted average of the particles.

PFs are particularly well-suited for battery systems, as they can handle the nonlinearities and uncertainties associated with battery dynamics, such as temperature effects, varying charge/discharge rates, and aging processes [9]. Furthermore, PFs can be used to estimate complex battery parameters, such as SOC and SOH, in real-time under uncertain operating conditions. Novel approaches such as particle flow filter frameworks with grey models have been developed to further enhance the performance of PFs for RUL prediction [14].

1.4.3 Robust Filters

Robust filters are designed to address the limitations of traditional filters when the noise characteristics are unknown or highly uncertain. In battery systems, noise can often be non-Gaussian and time-varying, making traditional filtering methods less effective. Zonotopic filtering, a robust filtering technique, uses geometric structures called zonotopes to represent uncertainty in the system's state.

Zonotopes are convex polytopes that provide a way to represent sets of possible states and measurement errors, making them useful for handling uncertainty in both the state and measurement models. This method is particularly advantageous when dealing with non-Gaussian noise and systems with bounded uncertainty, as it can efficiently estimate the state while taking into account the shape of the uncertainty set.

Zonotopic filters can be combined with other estimation techniques, such as Kalman filtering or particle filtering, to form hybrid filters that are both robust and efficient. These hybrid methods can be tailored to the specific needs of battery systems, where noise is often unknown and time-varying. Moreover, zonotopic filtering can be used in conjunction with particle swarm optimization (PSO) to further improve the robustness of the filter by adapting the filter's parameters in response to changes in the system's behavior.

1.4.4 The Integration of Filters with AI

The integration of traditional filter design methods with AI techniques is an emerging trend in the estimation of battery operating states. AI methods, such as DL and reinforcement learning (RL), are particularly suited for handling the large amounts of data generated by battery systems. By combining AI with filtering techniques, it is possible to improve the accuracy and robustness of state estimation, particularly in complex systems where the dynamics are not well understood or difficult to model.

For example, LSTM networks, a type of recurrent neural network, have been used to improve the estimation of SOC and SOH by learning the temporal dependencies in the battery data. Similarly, domain-adapted frameworks such as HybridoNet-Adapt combine DL with filtering techniques to enhance the accuracy of RUL prediction across different operating conditions [15]. RL techniques can be used to optimize filter parameters and adapt the filter's behavior in real-time based on the changing operating conditions of the battery system. By leveraging the strengths of both filtering and AI methods, it is possible to achieve more accurate and adaptive state estimation in battery systems.

Reviews of Bayesian-filtering-based techniques in RUL prediction for LIBs provide comprehensive insights into the integration of filtering methods with AI strategies [16]. These approaches demonstrate how the combination of traditional filtering principles with modern AI techniques can address the complex challenges of battery state estimation.

In conclusion, the design of effective filters for battery systems requires an understanding of both traditional estimation techniques and modern approaches that address the complexities of battery behavior. Linear, nonlinear, and robust filters each offer unique advantages depending on the nature of the system and the type of noise present. The integration of filtering methods with AI techniques further enhances the capability to accurately estimate key battery parameters, providing a comprehensive solution to the challenges of battery operating status analysis.

1.5 Artificial Intelligence Strategy

AI has become a key strategy in advancing the capabilities of battery systems, particularly in the fields of manufacturing, state estimation, and predictive maintenance [8]. The rapid development of AI techniques, including ML and DL, has introduced powerful tools for improving battery performance, enhancing the accuracy of SOC and SOH estimations, and optimizing the design and operation of battery systems.

In battery manufacturing, AI techniques are employed to optimize production processes, improve consistency, and predict manufacturing capacity. ML models can process large volumes of sensor data from production lines, identifying patterns that are not immediately apparent to human operators. These models are capable of making real-time adjustments to the manufacturing process, ensuring that production goals are met while maintaining high quality and efficiency. For instance, AI models based on supervised learning techniques, such as regression analysis and classification, can predict production outcomes based on input parameters, thus enhancing control over the manufacturing environment and reducing waste.

For modern battery systems, AI has become an indispensable tool for state estimation. Inferring the SOC, SOH, and RUL requires capturing strongly nonlinear and time-varying behaviors that conventional modeling approaches struggle to represent accurately. DL algorithms, particularly LSTM networks, are capable of learning temporal dependencies in battery data and can be trained to predict battery performance under various operating conditions [10]. PatchFormer, a novel patch-based transformer, has been developed for accurate RUL prediction of LIBs, demonstrating the potential of advanced AI architectures in this domain [13]. These models are capable of handling large amounts of data generated by battery systems and provide real-time updates of the state estimation, significantly improving the accuracy of predictions compared to traditional methods.

Furthermore, AI strategies enable predictive maintenance by providing advanced capabilities for forecasting battery degradation and predicting failure points. By analyzing historical data on battery performance, AI models can identify early signs of deterioration and predict when maintenance or replacement will be required. Survival analysis with ML has emerged as a particularly powerful approach for predicting Li-ion battery RUL [11]. This is particularly valuable in applications where battery failure can result in significant operational downtime or safety concerns, such as in EVs or renewable energy storage systems.

Another application of AI in battery systems is in the optimization of BMSs. AI techniques, including RL, can optimize the charging and discharging cycles of batteries to maximize their efficiency and lifespan. By continuously learning from the battery's behavior, RL algorithms can adapt to changes in the battery's operating conditions and adjust the control strategy accordingly. This leads to improved battery utilization and extended operational life, especially in systems with multiple batteries connected in series or parallel.

Large language models are also being explored for battery prognostics, opening up new possibilities for AI applications in battery health management. The integration of stress-informed transfer learning approaches enables accelerated battery life evaluation across diverse operating conditions and cell mechanisms, further expanding the capabilities of AI in battery systems.

In conclusion, AI has proven to be an invaluable strategy in improving the performance, reliability, and efficiency of battery systems. Through its application in manufacturing, state estimation, predictive maintenance, and optimization, AI provides the necessary tools to address the challenges posed by modern battery systems. As AI techniques continue to evolve, their integration into battery management and manufacturing systems will undoubtedly play a key role in driving the future of energy storage technologies.

1.6 Outlines

Although some progress has been made in the theory, methodology, and application of filtering techniques in battery systems, there is still much to be done. This book builds on the existing research, focusing on improving current methods and integrating them with emerging technologies. The content of this book highlights the continued development and application of state-of-the-art filtering methods in battery manufacturing, operating status analysis, and fault diagnosis.

The overall layout of the book, including its principal contents and their order, is summarized as follows: In Chapter 2, we present methods for optimizing battery manufacturing processes and predicting manufacturing capacity using data-driven models. Chapter 3 focuses on advanced state estimation techniques for battery systems, including methods for forecasting battery health and performance under various operational conditions. In Chapter 4, we explore the integration of advanced filtering methods with AI techniques for battery state estimation, emphasizing their effectiveness in nonstationary environments. In Chapter 5, we introduce advanced SOC estimation techniques using set-membership filtering methods and optimization strategies. These methods address the challenges of accurately estimating battery state under bounded uncertainties. Chapter 6 introduces methods for predicting the RUL of batteries, integrating advanced filtering strategies with AI to enhance the accuracy and reliability of RUL predictions. Finally, in Chapter 7, we summarize the research presented in this book and outline the future directions for the development of battery systems, focusing on the role of filtering techniques and AI in advancing battery technologies.

2

Advances in Battery Grouping by Artificial Intelligence Strategy

2.1 Introduction to Battery Grouping

Battery grouping is a critical process in battery manufacturing, particularly for ensuring the consistent performance of battery packs used in applications such as electric vehicles (EVs) and large-scale energy storage systems. Grouping refers to the classification of individual battery cells based on key performance parameters such as capacity, internal resistance, and voltage. By grouping cells with similar characteristics, manufacturers can ensure that the battery pack operates efficiently, safely, and with a longer lifespan [17]. Traditional methods of battery grouping often rely on empirical rules or manual sorting, which may lead to inconsistent performance and suboptimal grouping. These methods are time-consuming and may fail to account for the complex interdependencies between various battery parameters [18]. To address these challenges, more sophisticated techniques, such as clustering algorithms, have been applied to automate the grouping process and improve accuracy [19].

In recent years, optimization-based clustering methods, particularly those using genetic algorithms (GA), have gained attention for their ability to efficiently group batteries based on multiple parameters [12]. However, conventional GAs face several limitations, such as low search efficiency and the risk of premature convergence. To overcome these drawbacks, novel approaches have been developed, such as the chaotic dual-population genetic algorithm (CDP-GA), which introduces primary and secondary populations to enhance search accuracy and convergence speed. This chapter introduces the principles of battery grouping, discusses the limitations of traditional methods, and presents the application of advanced clustering techniques, including the CDP-GA algorithm, for optimizing battery grouping in manufacturing processes. The chapter highlights the importance of minimizing differences in key battery parameters within a group, as well as ensuring the consistency of battery packs to maximize performance and lifespan [20].

2.2 Problem Formulation and Model Establishment

2.2.1 Problem Formulation

Consider a set of objects characterized by a series of technical parameters and their corresponding weights. The objective of this study is to partition these objects into several clusters according to given requirements, such that the similarity among objects within the same cluster is maximized while the clustering structure remains interpretable and computationally efficient.

The following sections begin with a discussion of the underlying classification principles, upon which we formulate fast classification criteria adapted to the considered application setting.

2.2.2 Model Establishment

To develop fast classification criteria, the following three requirements should be simultaneously taken into account:

1. the overlap between different categories should be minimized;
2. the dispersion among samples within each category should be minimized;
3. the spatial extent occupied by each category should be consistent with the probability of containing samples.

Therefore, the construction of fast classification criteria can be formulated as a complex optimization problem, in which the objective function must comprehensively reflect the aforementioned three aspects.

By incorporating these criteria, the mathematical model is expressed as:

$$\begin{cases} F(b,z) = F_1 + F_2 + F_3, \\ F_1 = \sum_{i=1}^{Z-1} \sum_{j=i+1}^{Z} M_{ij}, \\ F_2 = \sum_{z=1}^{Z} \sqrt{\sum_{i=1}^{n_z} \sum_{j=1}^{m} \left| b_{ij} - \bar{b}_j \right|^2}, \\ F_3 = \sum_{i=1}^{Z-1} \sum_{j=i+1}^{Z} P_{ij}, \end{cases} \tag{2.1}$$

where M_{ij} denotes the overlap between the minimum inclusion space of the i-th category and that of the j-th category in the normalized feature space. The variable b_{ij} is the j-th normalized feature value of the i-th classified object, and $\bar{b}_j$ is the average value of the j-th feature over all classified objects after normalization. The scalar Z denotes the total number of categories, n_z is the number of objects in the z-th category, and P_{ij} quantifies the difference in density between the i-th and j-th categories.

On the basis of the aforementioned model, a clustering scheme is then designed and implemented by means of a genetic-algorithm-based optimization procedure.

In this section, a CDP-GA is developed by specifying the chromosome encoding scheme, the crossover strategies, the chaotic mutation mechanism, and the natural selection procedure.

2.3 Chromosome Encoding

In the proposed CDP-GA, an integer-encoding scheme is adopted, where the admissible integer range is determined by the prescribed number of classification categories. Each population contains n chromosomes. The coding sequence of the i-th chromosome in the main population at the k-th iteration is denoted by $C_i^0(k) = [c_{i1}^0(k), c_{i2}^0(k), \dots, c_{in}^0(k)]$, and the coding sequence of the i-th chromosome in the secondary population at the k-th iteration is given by $C_i^1(k) = [c_{i1}^1(k), c_{i2}^1(k), \dots, c_{in}^1(k)]$. The chromosome length is denoted by s, and each gene position on the chromosome corresponds to the category assignment of a specific object. The fitness function is directly taken as $F(b, z)$ defined in (2.1).

During the initialization of the CDP-GA populations, a tent chaotic map is employed to generate each chromosome [21]:

$$
\begin{aligned}
&c_i(0) = \mathrm{ceil}(z_j \times Z), \quad j \in [1, s], \quad i \in [1, N], \\
&z_{x+1} = \begin{cases} \mathrm{rand}, & x = 0, \\ \dfrac{z_x}{p}, & x \neq 0,\ z < p, \\ \dfrac{1 - z_x}{1 - p}, & x \neq 0,\ z \geq p, \end{cases}
\end{aligned}
\tag{2.2}
$$

where N denotes the number of individuals in each population, ceil($\cdot$) is the ceiling function, rand is a random number uniformly distributed in (0, 1), and p is the chaotic control parameter.

2.4 Crossover Modes Designing

Two distinct crossover modes are designed in CDP-GA: an inter-population crossover used in the early stages of the evolution and a non-inbreeding crossover applied in the later stages. The crossover scheme for the secondary population follows the conventional GA. The inter-population crossover serves as an information-exchange mechanism between the two populations, where individuals in the main population randomly select one of the two populations as a crossover partner pool.

In the later evolutionary stage, the non-inbreeding crossover strategy focuses on promoting diversity among individuals by encouraging crossover between chromosomes with low similarity. The decision of whether a given individual participates in crossover is made by comparing a crossover operator a with a prescribed upper bound of the crossover probability, denoted by a_{cr}. This operator a is constructed using the objective function and can be adaptively adjusted at a global level. When the condition $a < a_{cr}$ is satisfied, the individual is allowed to undergo crossover:

$$a = \left(1 - \frac{k}{\text{iter}}\right) \times \frac{O_i(k)}{\min(O(k))} \times \text{rand}, \tag{2.3}$$

where iter is the maximum number of iterations, k is the current iteration index, $O_i(k)$ is the fitness value associated with the current chromosome, and $\min(O(k))$ is the fitness value of the best individual at iteration k.

The crossover operation is carried out as follows:

$$\begin{cases} C_i(k) = \left[c_{i1}(k), c_{i2}(k), \cdots, c_s(k)\right]^{\mathrm{T}}, \\ C_i'(k) = (E_1 - \alpha)\, C_i(k) + \alpha\, \tilde{C}_i(k), \quad i \in [1, n], \\ \alpha = \operatorname{diag}\left[\alpha_1, \alpha_2, \cdots, \alpha_s\right]_{s\times s}, \\ I_1'(k) = \left[C_1'(k), \cdots, C_n'(k)\right], \end{cases} \tag{2.4}$$

where α is a random crossover matrix whose diagonal elements are drawn from $\{0, 1\}$, s is both the chromosome length and the number of objects, $C_i'(k)$ denotes the chromosome $C_i(k)$ after crossover, E_1 is the identity matrix, and $I_1'(k)$ denotes the newly generated population. The subpopulation $I_2(k)$ performs crossover according to the traditional GA procedure.

2.5 Population Mutations

For the main population, two mutation strategies are employed: a chaotic mutation applied in the early and late stages of evolution, and a minimum-similarity mutation adopted in the middle stage.

The basic idea of chaotic mutation is to replace conventional random numbers with a chaotic map for pointwise mutation, thereby constructing a chaotic mutation matrix $C(k)$:

$$\begin{aligned} C(k) &= \operatorname{diag}\left[\operatorname{ceil}(Z \times z_j)\right]_{s\times 1}, \quad j \in (1, s), \\ z_{x+1} &= \begin{cases} \text{rand}, & x = 0, \\ \dfrac{z_x}{p}, & x \neq 0,\ z < p, \\ \dfrac{1 - z_x}{1 - p}, & x \neq 0,\ z \geq p, \end{cases} \end{aligned} \tag{2.5}$$

where Z is the number of categories, s is the chromosome length, p is the chaotic control parameter, ceil($\cdot$) denotes the ceiling function, and rand generates a uniform random number in (0, 1).

The mutation operator b plays a role analogous to that of the crossover operator. It is compared with a mutation threshold (denoted by mut) to determine whether the mutation is activated:

$$b = \left(1.5 - \frac{k}{\text{iter}}\right) \times \frac{\min\left(O(k)\right)}{O_i(k)} \times \text{rand}. \tag{2.6}$$

Here, iter denotes the maximum number of iterations, k indexes the current iteration, $O_i(k)$ represents the fitness of the i-th chromosome at iteration k, and $\min(O(k))$ is the best fitness value achieved in the current generation.

Let β be the mutation transformation matrix. When $k < \text{iter}/2$ or $k > 3\,\text{iter}/4$, the chaotic mutation strategy is applied:

$$\begin{cases} C_i'(k) = \left[c_{i1}'(k), c_{i2}'(k), \dots, c_{is}'(k)\right]^{\mathrm{T}}, \\ C_i''(k) = (E_1 - \beta)\, C_i'(k) + \beta\, C(k), \\ \beta = \operatorname{diag}\left[\beta_1, \beta_2, \dots, \beta_s\right]_{s\times s}, \\ I_1''(k) = \left[C_1''(k), \dots, C_n''(k)\right], \end{cases} \tag{2.7}$$

where $C_i''(k)$ is the mutated chromosome, E_1 is the identity matrix, and $I_1''(k)$ denotes the population after mutation.

When $k \in [\text{iter}/2, 3\,\text{iter}/4]$, the minimum-similarity mutation strategy is instead adopted:

$$\begin{cases} C_i'(k) = \left[c_{i1}'(k), c_{i2}'(k), \cdots, c_{is}'(k)\right]^{\mathrm{T}}, \\ C_i''(k) = (E_1 - \beta)\, C_i'(k) + \beta\, \widetilde{C}_i(k), \\ \beta = \operatorname{diag}\left[\beta_1, \beta_2, \cdots, \beta_s\right]_{s\times s}, \\ \widetilde{C}_i(k) = \left[\tilde{c}_{i1}(k), \tilde{c}_{i2}(k), \cdots, \tilde{c}_{is}(k)\right]^{\mathrm{T}}, \\ I_1''(k) = \left[C_1''(k), \cdots, C_n''(k)\right], \end{cases} \tag{2.8}$$

where $\widetilde{C}_i(k)$ is constrained to differ from $C_i'(k)$ at a prescribed number of gene positions; by default, at least half of the gene positions are inconsistent, thereby enforcing a minimum degree of dissimilarity.

The natural selection mechanism of CDP-GA is implemented in a random manner, such that each individual has the same probability of being passed to the next generation. After natural selection, a chaotic micromutation with a very low mutation probability is applied to further enhance population diversity. The corresponding state-transition matrix for natural selection is denoted by Q, whereas the state-transition matrix for micromutation is denoted by W. The final step consists of a population-boosting operation: the most elite chromosome(s) in the subpopulation are swapped with the least elite chromosome(s) in the main population. If this boosting step is not executed, an elitism strategy is employed for the main population. The associated state-transition matrix of this boosting (or elitism) step is denoted by Y. Consequently, the evolution of the main population can be summarized as:

$$I_1(k+1) = Q_{ss} \cdot W_{ss} \cdot Y_{ss} \cdot I_1''(k), \tag{2.9}$$

where $I_1(k+1)$ denotes the updated main population at iteration $k+1$.

2.6 Convergence Analysis

The convergence analysis of the CDP-GA follows the same general framework as traditional GAs. Let the crossover operator be represented by the matrix H, the mutation operator by M, the natural selection operator by G, and the micromutation operator by M_1. The overall state transition induced by one generation of CDP-GA can then be expressed as $P_1 = H \cdot M \cdot G \cdot M_1$.

Let V denote the finite state space of all possible population chromosomes, independent of the specific encoding scheme. The population configuration at each iteration is regarded as a state in V. Let $Z(k)$ represent the best individual in the population at iteration k, and let f^* represent the globally optimal solution associated with the considered optimization problem. In this formulation, crossover, mutation, micromutation, and natural selection correspond to random transformations that map the current state to a new state. The state-transition matrices associated with H, M, G, and M_1 depend only on the current state and are time-invariant. Therefore, according to the Markov property, H, M, G, and M_1 are treated as random matrices. Additionally, the new state-transition matrices M and M_1 are entrywise positive random matrices, while G is a column-permissible random matrix.

Theorem 2.1 *Given a random matrix H, if M and M_1 are entrywise positive random matrices and G is a column-permissible random matrix, then $P_1 = H \cdot M \cdot G \cdot M_1$ is an entrywise positive matrix.*

Proof 2.1. Denote the (i, j)-th elements of H, M, G, and M_1 by h_{ij}, m_{ij}, g_{ij}, and m_{1ij}, respectively. Since M and M_1 are entrywise positive random matrices, the product $X = H \cdot M$ is a positive matrix. Furthermore, according to [22], the matrix $X = H \cdot M$ is a positive matrix in the sense that all its entries are strictly positive.

Next, consider a positive matrix X and a column-permissible random matrix G. Let $Y = X \cdot G$. Then Y is also a positive matrix. Specifically,

$$\forall i, z, \quad x_{iz} > 0, \tag{2.10}$$

and, since G is column-permissible, one has

$$\forall j, \exists z \text{ such that } g_{zj} > 0 \quad \Rightarrow \quad \forall i, j, \ y_{ij} = \sum_{z=1}^{N} x_{iz} g_{zj} > 0, \tag{2.11}$$

which implies that Y is a positive matrix and $P_1 = Y \cdot M_1$.

Since Y is positive,

$$\forall i, z, \quad y_{iz} > 0, \tag{2.12}$$

and M_1 is an entrywise positive random matrix,

$$m_{1ij} > 0, \quad i, j \in [1, N], \tag{2.13}$$

it follows that

$$\forall i, z, \ y_{iz} > 0, \quad \forall z, j, \ m_{1zj} > 0 \quad \Rightarrow \quad \forall i, j, \ p_{ij} = \sum_{z=1}^{N} y_{iz} m_{1zj} > 0. \tag{2.14}$$

Therefore, $P_1 = Y \cdot M_1 = H \cdot M \cdot G \cdot M_1$ is an entrywise positive matrix.

Theorem 2.2 *If the population boosting operation updates the main chromosome population by exchanging elite individuals, then the CDP-GA converges to the optimal solution.*

Proof 2.2. Consider Goldberg's model in [23] for analyzing the convergence of a single-population evolutionary algorithm:

$$y = u + v \ln N, \quad (u \geq 0, \ v > 0), \tag{2.15}$$

where N denotes the population size, and u, v are constants determined by the specific optimization problem.

In CDP-GA, extending (2.15) to the dual-population setting, let N denote the total population size, with $N = \sum_{i=1}^{2} N_i$, where N_i is the size of the i-th subpopulation ($i = 1, 2$). Suppose the overall problem of establishing the fast grouping standard is denoted by O. In view of (2.1), O can be decomposed into three subproblems O_1, O_2, and O_3. The constants u and v in (2.15) are determined by the fast grouping problem O. Accordingly, (2.15) can be generalized to the dual-population case as:

$$y' = \sum_{i=1}^{2} (u_i + v_i \ln N_i), \quad (u_i \geq 0, \ v_i > 0), \tag{2.16}$$

where u_i and v_i are determined by the subproblems associated with the i-th population.

Let the search space of the fast grouping problem O be $R = [R_1, R_2]$, where R_1 and R_2 are the search spaces corresponding to the main and secondary populations, respectively. It is clear that $||R_i||_1 \leq ||R||_1$, $i = 1, 2$. Moreover, since solving O is more complex than solving any of the subproblems O_1, O_2, or O_3, one has

$$u_i \geq u, \quad v_i > v, \quad i = 1, 2. \tag{2.17}$$

Combining (2.15), (2.16), and (2.17), it follows that

$$y' \geq ku + v \sum_{i=1}^{2} \ln N_i > ku + v \ln \sum_{i=1}^{2} N_i = ku + (y - u), \tag{2.18}$$

where $k > 1$ denotes the number of subproblems (here $k = 2$). From (2.18), it is obtained that $y' > y$, which implies that the convergence behavior of the dual-population algorithm model is superior to that of a single-population GA with the same total population size. Consequently, the CDP-GA converges to the optimal solution with probability 1.

Remark 2.1 When the algorithm is restricted to an elitist preservation strategy, it can also converge to the global optimal solution.

Since P_1 is a random matrix with all entries strictly positive, every non-global-optimal solution has a convergence probability of 0. Hence, the probability of the population being in the global optimum converges to 1, that is, as $k \to \infty$, one has $P\{Z(k) = f^*\} \to 1$.

2.7 Case Study on Battery Grouping

In this section, the proposed CDP-GA algorithm is further examined through a power battery grouping case study, which serves to verify its effectiveness. The experimental data were collected from the production line of Tianneng Battery Group (Zhejiang Province, China), and consist of capacity, internal resistance, and voltage measurements from 228 cells. The corresponding assembly line layout is illustrated in Figure 2.1.

The fast grouping standard obtained by the CDP-GA algorithm is depicted in Figure 2.2. From the results, it is apparent that the proposed method partitions the battery-indicator data space into three subspaces, each associated with one grouping standard that can be directly used for pack formation. For batteries located in overlapping regions of these subspaces, multiple standards may be applied, providing additional flexibility for practical grouping decisions.

Figure 2.1 Battery assembly line.

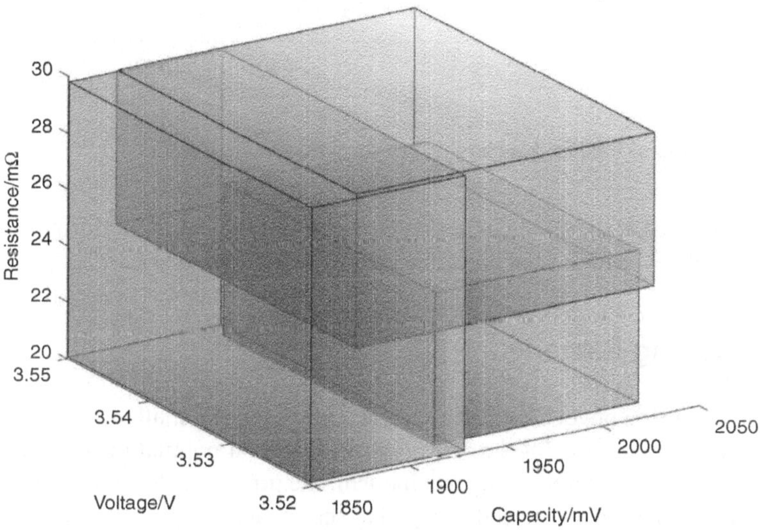

Figure 2.2 Standard rendering for fast grouping based on CDP-GA.

Figure 2.3 further compares the proposed CDP-GA with several benchmark algorithms: the conventional GA, a dual-population genetic algorithm (DGA) that incorporates inter-population crossover and population boosting, a chaotic genetic algorithm (CGA) that employs chaotic mapping in both population initialization and mutation, and an improved genetic algorithm (IP-GA) that modifies the initialization procedure and introduces adaptive mutation as in [24].

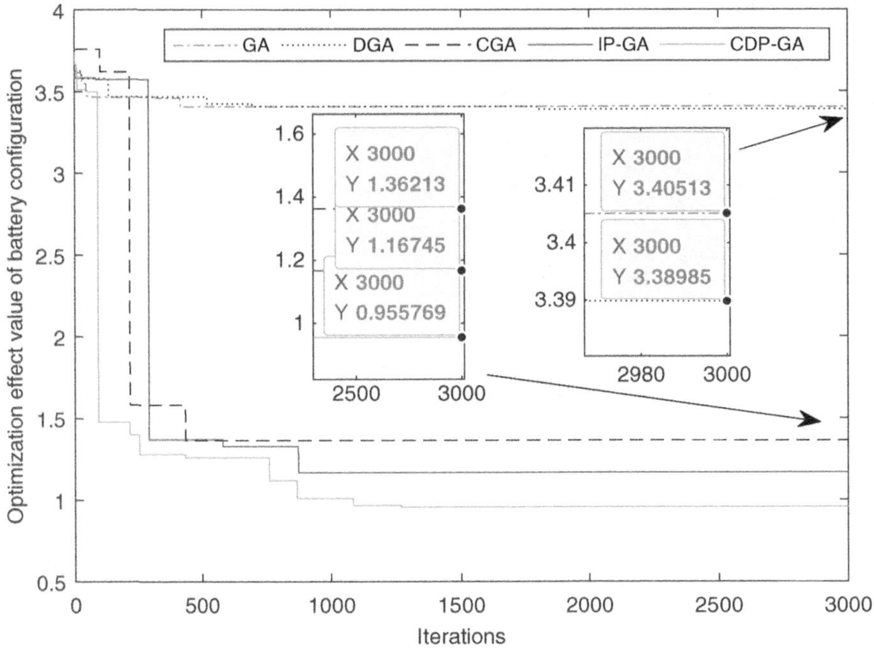

Figure 2.3 Rendering of the fast grouping results for different algorithms.

As shown in Figure 2.3, the CDP-GA achieves the best optimization performance among all candidates, with a final objective value of 0.955769.

2.8 Concluding Remarks

This chapter presents a novel CDP-GA aimed at addressing the challenge of rapid battery grouping. By introducing a primary–secondary dual-population structure combined with a population boosting mechanism, the proposed algorithm significantly enhances convergence speed and global search accuracy. CDP-GA effectively mitigates the common issues of premature convergence and low search efficiency that are typically associated with traditional GAs. The results from applying the algorithm to real-world power battery grouping tasks highlight its superior performance, demonstrating faster convergence and greater accuracy in global search compared to conventional methods. Ultimately, the proposed CDP-GA not only improves the grouping process efficiency but also offers a promising solution for optimizing the battery grouping process in the context of battery manufacturing and energy storage systems.

3

Advances in Forecasting of Battery Manufacturing Capacity by Artificial Intelligence Strategy

3.1 Introduction to Battery Manufacturing Capacity

In recent years, lithium-ion batteries have become one of the core components of new-energy vehicles and large-scale energy-storage systems. With the rapid expansion of the electric vehicle (EV) market and the continuous construction of large "gigafactories," battery manufacturers face growing pressure to ensure both high product quality and stable manufacturing capacity. According to the IEA's *Global EV Outlook 2024*, global EV battery demand has continued to grow rapidly in recent years [25]. Existing research on lithium-ion batteries has mainly focused on product-level quality analysis, such as state-of-charge (SOC) estimation, state-of-health (SOH) assessment, and remaining useful life prediction. These studies emphasize electrical performance and degradation behaviors from the perspective of a single cell or a battery pack. In contrast, the prediction and analysis of battery manufacturing capacity at the production-line level has not been extensively investigated, despite its crucial importance for production planning and strategic market deployment.

From the perspective of manufacturing systems, the capacity of a battery production line can be described as the number of qualified batteries produced per unit time under given process and resource constraints. This definition not only reflects the physical throughput of the line, but also implicitly embeds quality considerations through the notion of "qualified" products. In practical production, battery manufacturing capacity is influenced by multiple factors, including the performance of key equipment, process parameter settings, workforce and maintenance schedules, incoming material quality, and even external disturbances such as order fluctuations and supply-chain disruptions. Reasonable forecasting of manufacturing capacity, therefore, plays a central role in production planning, inventory management, and strategic capacity decisions in many high-tech manufacturing industries [26, 27]. As a result, the observed manufacturing capacity data usually exhibit pronounced nonlinearity, nonstationarity, and multi-time-scale behavior, which pose considerable challenges for accurate capacity forecasting.

Because the manufacturing capacity data are collected in chronological order with a fixed sampling interval, forecasting the manufacturing capacity of lithium-ion batteries can be naturally formulated as a time-series forecasting problem [28, 29]. Classical time-series models, such as ARIMA and exponential smoothing, have been widely used in industrial demand forecasting and process monitoring [28, 29]. However, these traditional models often assume approximate linearity and weak nonstationarity, which are not fully compatible with the strongly nonlinear, mode-mixed, and noise-contaminated characteristics of modern battery manufacturing data. Consequently, directly applying such models to capacity forecasting may lead to limited accuracy and poor robustness in the presence of complex fluctuations.

To handle complex nonstationary sequences, decomposition–reconstruction strategies have attracted increasing attention. Variational mode decomposition (VMD) is a powerful technique that adaptively decomposes a complex series into a finite number of band-limited intrinsic mode functions (IMFs) by solving a constrained variational optimization problem [30, 31]. Compared with empirical mode decomposition and some traditional filtering methods, VMD offers better mathematical grounding, reduced mode mixing, and flexible control of frequency bandwidths. These properties make VMD particularly suitable for extracting hidden multi-scale patterns from manufacturing capacity data and converting the original nonstationary series into several more regular subseries that are easier to model and forecast. In recent industrial applications, VMD has been successfully combined with deep neural networks to forecast nonstationary electricity load and power demand series, achieving higher accuracy than single-model baselines [32, 33].

However, the prediction performance of VMD-based methods highly depends on the selection of key parameters, especially the number of decomposition layers K and the penalty factor η. These parameters influence the center frequencies and bandwidths of the IMFs, thus affecting the degree of mode separation and the trade-off between decomposition accuracy and computational complexity [34–36]. If K is set too large, spurious modes and mode aliasing may occur, leading to redundant components and decreased interpretability. If η is set too small, each IMF becomes overly narrowband, which helps reduce mode aliasing but significantly increases the time cost of the decomposition. In contrast, other parameters such as the discrimination accuracy ε and noise tolerance τ generally have a less significant impact on the decomposition quality [35, 36]. Therefore, how to reasonably and automatically choose K and η for different manufacturing datasets is a key issue in practical applications.

After the manufacturing capacity series is decomposed into multiple IMFs, another important problem is how to construct accurate prediction models for each component and then reconstruct the final capacity forecast. In recent years, neural-network-based approaches have become widely adopted for time-series forecasting owing to their strong nonlinear mapping capability and data-driven learning nature [37, 38]. Among them, the LSTM network is particularly suitable for capturing long-range temporal dependencies and dealing with complex dynamic patterns. For example, [39] successfully applied an LSTM-based model to accurately forecast flood storage ponds, demonstrating its superiority over traditional models in handling nonlinear and nonstationary hydrological sequences.

Nevertheless, the performance of conventional LSTM models strongly depends on the manual selection of hyperparameters, such as the number of hidden-layer nodes, learning rate, and batch size [40, 41]. In many practical applications, these hyperparameters are determined by trial-and-error or empirical tuning, which makes the modeling process subjective and may lead to suboptimal results. In particular, the number of hidden nodes has a decisive influence on both the representation capacity and the generalization performance of the LSTM network. An insufficient number of hidden nodes may cause underfitting, whereas an excessively large number of nodes may increase the risk of overfitting and computational burden [40, 41].

Reinforcement learning, especially the Q-learning algorithm, provides a promising tool for data-driven decision optimization. By interacting with an environment and receiving scalar rewards, Q-learning adaptively updates a state–action value table and gradually learns an optimal decision policy without requiring an explicit model of the environment. Recent work has formulated hyperparameter tuning of neural networks as a Markov decision process (MDP) and used Q-learning to search for efficient hyperparameter configurations, achieving better performance than traditional search methods [42]. This property makes Q-learning attractive for hyperparameter optimization in machine-learning models, where the environment is the training process and the reward can be defined based on prediction errors or evaluation metrics. In parallel, deep reinforcement learning (RL) has also been applied to optimize capacity planning policies in complex manufacturing environments, showing clear advantages over heuristic rules in terms of throughput and cycle time [43].

Motivated by the aforementioned observations, this chapter focuses on the forecasting problem of lithium-ion battery manufacturing capacity with nonlinear and nonstationary characteristics. The central idea is to integrate VMD and LSTM within a reinforcement-learning-based optimization framework. First, an MDP is constructed to describe the parameter-selection problem of VMD, and a Q-learning strategy is designed to automatically search for the optimal number of decomposition layers K and penalty factor η. The original manufacturing capacity series is then decomposed using VMD with these optimized parameters, yielding IMFs that exhibit improved mode separation and smoother temporal patterns. Second, a new Q-learning-based method is developed to optimize the key hyperparameters of the LSTM model, particularly the number of hidden-layer nodes, thereby reducing the randomness and subjectivity of conventional LSTM configuration. Based on the optimally decomposed IMFs and the optimally tuned LSTM models, a Q-learning-based VMD and long short-term memory (Q-VMD-LSTM) forecasting algorithm is finally constructed to enhance the prediction accuracy of nonstationary lithium-ion battery manufacturing capacity.

3.2 Problem Formulation

Analysis of the original battery manufacturing capacity data shows that the corresponding time series exhibits pronounced nonstationary behavior. To mitigate the adverse impact of this nonstationarity on forecasting performance, the original sequence is first decomposed, and each resulting component is subsequently predicted. In this context, two key issues must be addressed:

1. how to optimally determine the number of decomposition layers and the penalty factor in the VMD method;
2. how to enhance the prediction accuracy of the decomposed components when using the LSTM algorithm.

Both issues can be reformulated as reinforcement-learning-based optimization tasks for forecasting nonstationary battery manufacturing capacity, and can thus be viewed as optimal parameter optimization problems.

For a given sequence $X(t)$, the marginal spectrum of each modal function $u_k(t)$ is obtained via the Hilbert transform, and the center frequency exponent w_k for each mode is determined accordingly. By taking the gradient in the sense of the Euclidean 2-norm, a constrained variational problem is derived. The bandwidths of the K modes are then obtained by solving

$$\begin{aligned} &\min_{\{u_k, w_k\}} \left\{ \sum_{k=1}^{K} \left\| \partial_t \left[\left(\delta(t) + \frac{j}{\pi t} \right) * u_k(t) e^{-j w_k t} \right] \right\|_2^2 \right\} \\ &\text{s.t.} \ \sum_{k=1}^{K} u_k(t) = X(t), \end{aligned} \tag{3.1}$$

where u_k denotes the k-th modal component, w_k is its center frequency, the quantity $X(t)$ stands for the original signal, j stands for the imaginary unit, and $\delta(t)$ represents the Dirac delta function.

3.3 Q-learning Principle

RL provides a general framework for learning decision-making policies through interaction with the environment. In contrast to supervised learning, where labeled input–output pairs are given in advance, an RL agent must choose actions, observe the resulting state transitions and scalar rewards, and gradually improve its behavior based on long-term return. Q-learning is one of the most widely used value-based RL algorithms, and is particularly suitable for optimization problems with discrete state and action spaces [44].

3.3.1 Markov Decision Process

The RL framework is commonly formulated on the basis of an MDP. A finite MDP is represented by the tuple

$$\mathcal{M} = \langle \mathcal{S}, \mathcal{A}, \mathcal{P}, r, \gamma \rangle,$$

where $\mathcal{S}$ denotes the set of states, $\mathcal{A}$ denotes the set of actions, $\mathcal{P}(s' \mid s, a)$ specifies the probability of moving from state $s \in \mathcal{S}$ to state $s' \in \mathcal{S}$ when action $a \in \mathcal{A}$ is chosen, $r(s, a)$ gives the immediate reward received by taking action a in state s, and $\gamma \in [0, 1)$ is the discount factor that down-weights future rewards.

At each discrete time step t, the agent observes the current state $S_t \in \mathcal{S}$, selects an action $A_t \in \mathcal{A}$ according to a policy, obtains an instantaneous reward R_{t+1}, and the system evolves to the next state S_{t+1}. The aim of RL is to design a policy that maximizes the expected long-term discounted return.

A (stochastic) policy π maps each state to a probability distribution over actions, so that

$$\pi(a \mid s) = \Pr\{A_t = a \mid S_t = s\}.$$

Given a fixed policy π, the return starting from time t is defined as:

$$G_t = \sum_{k=0}^{\infty} \gamma^k R_{t+k+1}, \tag{3.2}$$

and the associated state-value function $v^\pi(s)$ and action-value function $q^\pi(s, a)$ are defined by

$$v^\pi(s) = \mathbb{E}_\pi [G_t \mid S_t = s], \tag{3.3}$$

$$q^\pi(s, a) = \mathbb{E}_\pi [G_t \mid S_t = s, A_t = a], \tag{3.4}$$

where the expectation is evaluated over the trajectories distribution generated by policy π under the MDP dynamics. These definitions coincide with the standard discounted MDP formulation used in RL.

3.3.2 Bellman Optimality Equation and Q-function

Among all policies, an optimal policy π^* achieves the maximal expected return for every state. The corresponding optimal state-value function $v^*(s)$ and optimal action-value function $q^*(s, a)$ are defined as:

$$v^*(s) = \max_{\pi} v^{\pi}(s), \tag{3.5}$$

$$q^*(s, a) = \max_{\pi} q^{\pi}(s, a). \tag{3.6}$$

These optimal value functions satisfy the Bellman optimality equations. In particular, the optimal action-value function $q^*(s, a)$ satisfies:

$$q^*(s, a) = \mathbb{E}\left[R_{t+1} + \gamma \max_{a' \in \mathcal{A}} q^*(S_{t+1}, a') \,\middle|\, S_t = s, A_t = a\right]. \tag{3.7}$$

Once $q^*(s, a)$ is known, an optimal policy can be obtained by selecting, at each state s, any action that maximizes $q^*(s, a)$:

$$\pi^*(s) \in \arg\max_{a \in \mathcal{A}} q^*(s, a). \tag{3.8}$$

Therefore, many RL algorithms aim at learning an approximation of q^* directly from interaction data.

3.3.3 Tabular Q-learning Algorithm

Q-learning belongs to the class of model-free RL algorithms and incrementally learns the optimal action-value function $q^*(s, a)$ using temporal-difference updates. In the tabular setting, a Q-table $Q(s, a)$ is maintained for all state–action pairs $(s, a) \in \mathcal{S} \times \mathcal{A}$. At each time step t, after executing action A_t in state S_t, observing reward R_{t+1} and the next state S_{t+1}, the Q-table is updated according to:

$$Q_{t+1}(S_t, A_t) = (1 - \alpha)Q_t(S_t, A_t) + \alpha\left[R_{t+1} + \gamma \max_{a' \in \mathcal{A}} Q_t(S_{t+1}, a')\right], \tag{3.9}$$

where $\alpha \in (0, 1]$ is the learning rate and γ is the discount factor. The term in brackets is the temporal-difference target, composed of the immediate reward and the discounted estimate of the optimal future value. The update in (3.9) can be interpreted as a stochastic approximation to the Bellman optimality Eq. (3.7), and is known as the Q-learning rule.

During learning, the agent must balance *exploration* and *exploitation*. A widely used strategy is the ε-greedy policy: with probability $1 - \varepsilon$, the agent selects the action with the highest current Q-value, $A_t = \arg\max_{a \in \mathcal{A}} Q_t(S_t, a)$, and with probability ε, it chooses an action uniformly at random from $\mathcal{A}$. The exploration rate ε is often decreased gradually as learning progresses, so that the policy becomes more deterministic when the Q-table has converged.

Under suitable technical conditions (finite state and action sets, sufficiently small but nonvanishing learning rate, and adequate exploration), tabular

Q-learning is guaranteed to converge to the optimal action-value function q^* with probability one. In practice, Q-learning is simple to implement and does not require explicit knowledge of the transition probabilities $\mathcal{P}$, which makes it attractive for data-driven optimization problems.

3.3.4 Formulation for Parameter Optimization

In this chapter, the Q-learning algorithm is employed as a discrete search mechanism for hyperparameter optimization. The selection of algorithmic parameters (such as the number of VMD decomposition layers and the LSTM hidden-layer size) is formulated as an MDP, where each state represents a current candidate parameter configuration, each action corresponds to adjusting one or several parameters, and the reward is defined based on the decomposition quality or prediction error. By iteratively updating the Q table according to (3.9) and following an ε-greedy exploration strategy, the learning agent can gradually identify parameter combinations that yield better forecasting performance. The detailed design of the state, action, and reward for the battery manufacturing capacity forecasting problem will be presented in Section 3.6.

3.4 Variational Mode Decomposition

As discussed in Section 3.1, the original manufacturing capacity sequence of a lithium-ion battery production line usually exhibits strong nonlinearity and non-stationarity. Directly building a forecasting model on such data often leads to poor accuracy. To alleviate this problem, this book adopts VMD to separate the original sequence into a finite number of IMFs with relatively narrow bandwidths before prediction [29, 30, 40, 41]. Each IMF mainly occupies a limited frequency band around a certain center frequency, which makes it easier to capture the underlying temporal patterns with subsequent learning models.

3.4.1 Constrained Variational Formulation

Consider an observed real-valued signal $X(t)$, which in this chapter represents the battery manufacturing capacity series. VMD assumes that $X(t)$ can be decomposed into K modes $u_k(t)$, each of which is an amplitude-modulated and frequency-modulated (AM–FM) component with center frequency ω_k. For a given set of modes $\{u_k(t)\}_{k=1}^{K}$ and center frequencies $\{\omega_k\}_{k=1}^{K}$, the analytic signal of each mode is obtained by applying the Hilbert transform, i.e.,

$$\left(\delta(t) + \frac{j}{\pi t}\right) * u_k(t),$$

where $\delta(t)$ is the Dirac delta and j is the imaginary unit. After frequency shifting by $e^{-j\omega_k t}$, VMD measures the bandwidth of each mode by the squared ℓ_2-norm of the gradient of the demodulated analytic signal. Summing over all modes, the constrained variational problem can be written as:

$$\begin{aligned} \min_{\{u_k\},\{\omega_k\}} \quad & \sum_{k=1}^{K} \left\| \partial_t \left[\left(\delta(t) + \frac{j}{\pi t} \right) * u_k(t) \right] e^{-j\omega_k t} \right\|_2^2 \\ \text{s.t.} \quad & \sum_{k=1}^{K} u_k(t) = X(t), \end{aligned} \tag{3.10}$$

where ∂_t denotes the time derivative. The constraint in (3.10) enforces that the sum of all modal components reconstructs the original signal.

To solve (3.10), a penalty factor $\eta > 0$ and a Lagrange multiplier $\lambda(t)$ are introduced, leading to the following augmented Lagrangian:

$$\begin{aligned} L(\{u_k\},\{\omega_k\},\lambda) = {} & \eta \sum_{k=1}^{K} \left\| \partial_t \left[\left(\delta(t) + \frac{j}{\pi t} \right) * u_k(t) \right] e^{-j\omega_k t} \right\|_2^2 \\ & + \left\| X(t) - \sum_{k=1}^{K} u_k(t) \right\|_2^2 + \left\langle \lambda(t), X(t) - \sum_{k=1}^{K} u_k(t) \right\rangle, \end{aligned} \tag{3.11}$$

where $\langle \cdot, \cdot \rangle$ denotes the inner product. The penalty factor η controls the balance between narrowband enforcement and reconstruction fidelity, and its value has a significant influence on the decomposition result [40, 41].

3.4.2 Iterative Solution by ADMM

The augmented Lagrangian (3.11) is minimized by an alternating direction method of multipliers (ADMM) scheme. In practice, it is convenient to work in the frequency domain. Let $\hat{f}(\omega)$, $\hat{u}_k(\omega)$, and $\hat{\lambda}(\omega)$ denote the Fourier transforms of $X(t)$, $u_k(t)$, and $\lambda(t)$, respectively, and let the superscript n denote the iteration index. The updates of the modal components, center frequencies, and Lagrange multiplier are given by

$$\hat{u}_k^{n+1}(\omega) = \frac{\hat{f}(\omega) - \sum_{i \neq k} \hat{u}_i^n(\omega) + \frac{1}{2}\hat{\lambda}^n(\omega)}{1 + 2\eta\left(\omega - \omega_k^n\right)^2}, \tag{3.12}$$

$$\omega_k^{n+1} = \frac{\int_0^\infty \omega \left| \hat{u}_k^{n+1}(\omega) \right|^2 d\omega}{\int_0^\infty \left| \hat{u}_k^{n+1}(\omega) \right|^2 d\omega}, \tag{3.13}$$

$$\hat{\lambda}^{n+1}(\omega) = \hat{\lambda}^n(\omega) + \tau \left[\hat{f}(\omega) - \sum_{k=1}^{K} \hat{u}_k^{n+1}(\omega) \right], \tag{3.14}$$

where $\tau > 0$ is a step-size parameter for updating the Lagrange multiplier. Equations (3.12)–(3.14) are iteratively applied to all modes until convergence.

In numerical implementation, the convergence of the VMD iteration is monitored by the relative change of the modes between two successive iterations. Specifically, the iteration is terminated when

$$\sum_{k=1}^{K} \frac{\left\| \hat{u}_k^{n+1} - \hat{u}_k^{n} \right\|_2^2}{\left\| \hat{u}_k^{n} \right\|_2^2} < \varepsilon, \tag{3.15}$$

where $\varepsilon > 0$ is a small tolerance controlling the decomposition accuracy. Once (3.15) is satisfied, the current estimates $\{u_k(t)\}_{k=1}^{K}$ are taken as the final IMFs of the original signal.

3.4.3 Remarks on Parameter Selection

The VMD algorithm described earlier depends on several parameters, among which the number of modes K and penalty factor η are the most critical [40, 41]. If K is chosen too small, important oscillatory components may be merged; if K is excessively large, mode splitting and aliasing can occur. A small η yields very narrowband modes but increases computation time, whereas a large η may lead to insufficient separation. Other parameters, such as the noise-tolerance factor τ, initialization of the center frequencies, and convergence threshold ε, normally have a less pronounced impact on the decomposition result and can be set to commonly used default values.

In this chapter, the determination of K and η is treated as an optimization problem and will be carried out by a Q-learning-based search strategy. A decomposition-quality index, defined later in Section 3.6, is adopted as the optimization objective to automatically select an appropriate combination of K and η for the nonstationary battery manufacturing capacity sequence.

3.5 Long Short-term Memory

As discussed in Section 3.1, once the original manufacturing capacity series has been decomposed into several IMFs by VMD, it is necessary to build accurate forecasting models for each IMF and then reconstruct the final capacity prediction. Owing to their strong nonlinear mapping capability and ability to capture long-range temporal dependencies, LSTM networks have been widely adopted in time-series forecasting problems [37, 38]. For example, [39] demonstrated that an LSTM-based model can effectively forecast complex hydrological sequences, highlighting its suitability for nonlinear and nonstationary signals.

An LSTM network can be interpreted as a particular form of recurrent neural network that incorporates gating mechanisms. At each time step t, given the input vector $\boldsymbol{x}_t$, the previous hidden state $\boldsymbol{h}_{t-1}$, and the previous cell state $\mathbf{c}_{t-1}$, the LSTM cell computes

$$\begin{aligned}
\boldsymbol{f}_t &= \sigma(W_f \boldsymbol{x}_t + U_f \boldsymbol{h}_{t-1} + \boldsymbol{b}_f),\\
\boldsymbol{i}_t &= \sigma(W_i \boldsymbol{x}_t + U_i \boldsymbol{h}_{t-1} + \boldsymbol{b}_i),\\
\tilde{\mathbf{c}}_t &= \tanh(W_c \boldsymbol{x}_t + U_c \boldsymbol{h}_{t-1} + \boldsymbol{b}_c),\\
\mathbf{c}_t &= \boldsymbol{f}_t \odot \mathbf{c}_{t-1} + \boldsymbol{i}_t \odot \tilde{\mathbf{c}}_t,\\
\mathbf{o}_t &= \sigma(W_o \boldsymbol{x}_t + U_o \boldsymbol{h}_{t-1} + \boldsymbol{b}_o),\\
\boldsymbol{h}_t &= \mathbf{o}_t \odot \tanh(\mathbf{c}_t),
\end{aligned} \tag{3.16}$$

where $\boldsymbol{f}_t$, $\boldsymbol{i}_t$, and $\mathbf{o}_t$ denote the forget, input, and output gates, respectively; $\tilde{\mathbf{c}}_t$ is the candidate cell state; $W_.$ and $U_.$ are weight matrices; $\boldsymbol{b}_.$ are bias vectors; $\odot$ stands for element-wise multiplication; the functions $\sigma(\cdot)$ and $\tanh(\cdot)$ correspond to the sigmoid and hyperbolic tangent functions, respectively. Through these gated operations, the LSTM cell can selectively retain or discard historical information, which alleviates the gradient-vanishing problem and enhances its capability to model long-term dependencies.

In the context of battery manufacturing capacity forecasting, each IMF obtained from VMD is treated as a relatively regular subseries. For each IMF, an LSTM-based forecasting model is constructed by feeding a sliding window of past samples as the input sequence and predicting the next (or multiple future) values as the output. The hidden-layer state dimension, learning rate, and other hyperparameters significantly affect the representation ability and generalization performance of the LSTM network [40, 41]. In particular, too few hidden units may lead to underfitting, whereas too many hidden units may increase overfitting risk and computational burden. To reduce the subjectivity of manual hyperparameter tuning, this chapter employs the Q-learning strategy introduced in Section 3.3 to adaptively search for suitable LSTM configurations. The detailed design of the Q-learning-based VMD-LSTM forecasting framework will be presented in Section 3.6.

3.6 Design of the Forecasting Method

We first reformulate Eq. (3.1) through the introduction of the penalty factor η together with the Lagrange multiplier $\lambda(t)$, leading to the augmented functional

$$\begin{aligned}
L(u_k, w_k, \lambda) = \eta \sum_{k=1}^{K} &\left\| \partial_t \left[\left(\delta(t) + \frac{j}{\pi t} \right) * u_k(t) e^{-j w_k t} \right] \right\|_2^2 \\
&+ \left\| X(t) - \sum_{k=1}^{K} u_k(t) \right\|_2^2 \\
&+ \left\langle \lambda(t), X(t) - \sum_{k=1}^{K} u_k(t) \right\rangle .
\end{aligned} \tag{3.17}$$

The updates of u_k and w_k in Eq. (3.17) are given by

$$\hat{u}_k^{n+1}(t) = \frac{\hat{f}(t) - \sum_{i=1}^{K} \hat{u}_i(t) + \frac{\hat{\lambda}(t)}{2}}{1 + 2\eta(t - t_k)^2}, \tag{3.18}$$

$$w_k^{n+1} = \frac{\int_0^\infty w|\hat{u}_k(t)|^2 dt}{\int_0^\infty |\hat{u}_k(t)|^2 dt}, \tag{3.19}$$

and the Lagrange multiplier is updated as:

$$\hat{\lambda}^{n+1}(t) = \hat{\lambda}^n(t) + \tau\left(\hat{f}(t) - \sum_{k=1}^{K} \hat{u}_k^{n+1}(t)\right), \tag{3.20}$$

where τ denotes the step size of the multiplier update.

Convergence of the decomposition is checked by verifying whether the following condition is satisfied:

$$\sum_{k=1}^{K} \frac{\left\|\hat{u}_k^{n+1} - \hat{u}_k^n\right\|_2^2}{\left\|\hat{u}_k^n\right\|_2^2} < \varepsilon, \tag{3.21}$$

in which case the iterative process terminates, and all modal components are output.

Under complex manufacturing conditions for lithium-ion batteries, the nonstationarity of the time-series data can significantly deteriorate prediction performance. To address this issue, a Q-learning-based RL scheme is adopted to optimize the key parameters of the VMD algorithm, thereby ensuring an appropriate decomposition of the original series. Specifically, the penalty factor η and the number of decomposition layers K are treated as decision variables to be optimized by the Q-learning algorithm, and the optimal parameter pair (η^*, K^*) is obtained through training. Based on this idea, a Q-VMD-LSTM framework is constructed, as illustrated in Figure 3.1.

Starting with data acquisition, the manufacturing capacity of the lithium-ion battery is first correlated with the qualification rate, production volume, and production time, and the raw data are preprocessed to form a suitable manufacturing capacity dataset. For datasets with pronounced nonstationary characteristics, the VMD algorithm is then employed to perform mode decomposition. The Q-learning algorithm is used to adaptively select the optimal values of the VMD key parameters, and the resulting optimized VMD scheme is applied to decompose the original sequence into multiple components. Each component is subsequently predicted by an LSTM model, whose key parameters are also tuned via Q-learning. Finally, the predicted components are recombined to reconstruct the forecast of the original sequence, and the proposed Q-VMD-LSTM algorithm is further evaluated through a simulation study designed to verify its effectiveness.

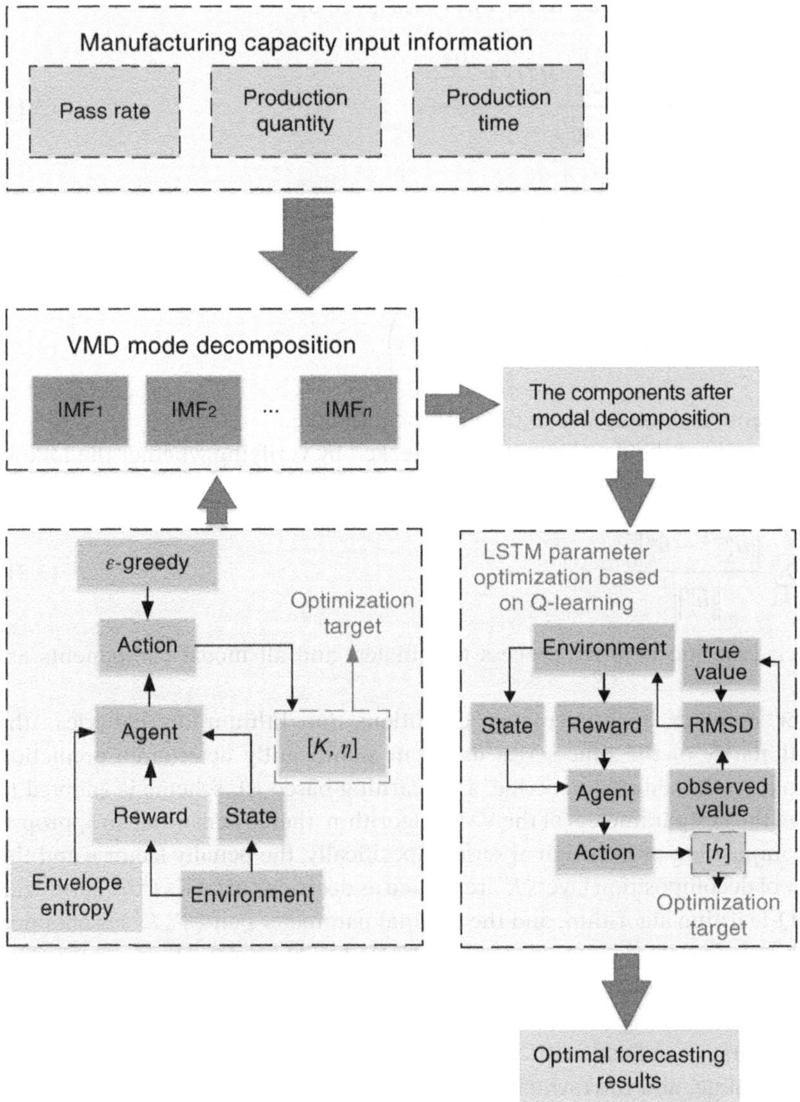

Figure 3.1 Structure of the Q-VMD-LSTM algorithm.

The operational procedure of the proposed Q-VMD-LSTM algorithm proceeds according to the following steps:

Step 1: Based on the qualified rate, production quantity, and production time collected from the production line, the manufacturing capacity of qualified batteries per unit time is used to construct the battery manufacturing capability index.

The original manufacturing capacity sequence $X(t)$ is defined as $X(t) = \frac{Q(t) \times R(t)}{T}$, where $Q(t)$ denotes the number of batteries produced per unit time, $R(t)$ denotes the production pass rate, and T is the corresponding time interval.

Step 2: The VMD algorithm is employed to process the lithium-ion battery sequence $X(t)$, which exhibits complex features such as nonstationarity, to obtain a modal decomposition. A Q-learning algorithm is then employed to search for the optimal parameter set (η, K) of the VMD algorithm, where η is the penalty factor and K is the number of decomposition modes.

In the Q-learning framework, an index must be specified as the optimization objective. In this study, the minimization of the VMD-based envelope entropy E_p is chosen as the optimization target. The envelope entropy reflects the uniformity of the probability distribution: the more concentrated the distribution, the smaller E_p. It is computed as:

$$\begin{cases} p_j = \dfrac{a(j)}{\sum_{j=1}^{N} a(j)}, \\ E_p = -\sum\limits_{j=1}^{N} p_j \lg p_j, \end{cases} \tag{3.22}$$

where $a(j)$ is the envelope signal obtained by demodulating $X(j)$ via the Hilbert transform, and p_j is the normalized envelope.

The optimization of the VMD parameter pair (η, K) is abstracted as an MDP, for which the state set S, action set A, loss function L, and reward function R are designed as follows. Since the objective is to determine the parameter combination that yields the most accurate VMD decomposition, the state matrix is defined as:

$$S_1 = [\eta, K]. \tag{3.23}$$

The agent seeks to minimize the envelope entropy, and the action matrix is chosen as:

$$A_1 = [\Delta K, \Delta \eta], \tag{3.24}$$

which represents the incremental adjustment of the state S_1 by fixed steps in K and η.

The loss function of the Q-learning algorithm is consistent with Eq. (3.22), since the goal is to minimize E_p:

$$L_1 = E_p = -\sum_{j=1}^{N} p_j \lg p_j. \tag{3.25}$$

To encourage the agent to explore state–action pairs that reduce the envelope entropy, the reward function is defined by

$$R_1 = \frac{1}{E_p} \times 100 = \frac{1}{-\sum_{j=1}^{N} p_j \lg p_j} \times 100. \tag{3.26}$$

During learning, the Q-learning agent continuously updates the Q-table by iterating over state–action pairs, and ultimately converges to the optimal action policy for each state according to the expected Q-values. The Q-value update rule is given by

$$Q(s, a) \leftarrow (1 - \alpha)\, Q(s, a) + \alpha\gamma \max_{a'} Q(s', a') + \alpha R_1,$$

where α is the learning rate, γ is the discount factor, s' is the next state, and a' is a candidate action in state s'.

After sufficient iterations, the agent yields a converged Q-table. By inspecting this table, the optimal action under each state (η, K) can be identified, and thus the optimal penalty factor η and number of decomposition layers K that minimize the envelope entropy are obtained. In this way, the VMD algorithm is endowed with an adaptive parameter-selection mechanism that achieves an improved decomposition effect.

Step 3: Using the optimized VMD parameters, the original sequence $X(t)$ is decomposed into K intrinsic components, $X^1(t),\ X^2(t),\ \ldots,\ X^K(t)$, each of which captures a different oscillatory mode of the manufacturing capacity series.

Step 4: A second Q-learning stage is used to optimize the structure of the LSTM predictor by tuning the number of hidden-layer nodes h. The optimization procedure is analogous to that in Step 2. Specifically, the state is chosen as $S_2 = [h]$, and the action matrix is defined as $A_2 = [-\Delta x,\ \Delta x]$, where Δx denotes the step size for adjusting the number of hidden nodes. The objective is to improve the predictive accuracy of the LSTM model, and thus the root-mean-square error (RMSE) of the LSTM prediction is adopted as the loss function:

$$L_2 = \sqrt{\frac{1}{T}\sum_{t=1}^{T} \left(y_t - \hat{y}_t\right)^2}, \tag{3.27}$$

where y_t and $\hat{y}_t$ denote the true and predicted values at time t, respectively, and T is the prediction horizon.

The corresponding reward function is defined by

$$R_2 = \frac{1}{L_2} = \frac{1}{\sqrt{\frac{1}{T}\sum_{t=1}^{T} \left(y_t - \hat{y}_t\right)^2}}, \tag{3.28}$$

so that smaller RMSE values lead to higher rewards. The Q-value update rule is then

$$Q(s,a) \leftarrow (1-\alpha)\,Q(s,a) + \alpha\gamma \max_{a'} Q(s',a') + \alpha R_2.$$

Through iterative updates, the optimal number of hidden-layer nodes h is obtained, and the corresponding LSTM model is adopted as the final predictor.

Step 5: The optimized LSTM model is used to forecast each decomposed component obtained in Step 3. The individual component forecasts are then superimposed to reconstruct the final prediction of the original manufacturing capacity sequence.

Step 6: Finally, three statistical indicators, RMSE, mean absolute error (MAE), and mean absolute percentage error (MAPE), are then calculated to assess the prediction performance. These metrics are used to assess the effectiveness of the proposed Q-VMD-LSTM algorithm for forecasting the nonstationary lithium battery manufacturing capacity.

3.7 Case Study on Forecasting of Battery Manufacturing Capacity

In this study, the production data collected from the 18650 battery workshop of Zhejiang Tianneng Battery Co., Ltd. are employed to assess the performance of the proposed algorithm. As this workshop is still in trial operation, the data transmission is not yet fully stable, and the resulting manufacturing capacity measurements exhibit pronounced nonstationarity.

3.7.1 Model Parameter Setting

Except for the key parameters—the penalty factor α and the number of decomposition modes K, which are determined adaptively by the Q-learning algorithm—the remaining VMD parameters are configured as summarized in Table 3.1. The three parameters listed are commonly used defaults and do not have a significant impact on the decomposition performance.

Table 3.1 VMD parameter settings.

Argument	Parameter meaning	Parameter value
τ	Noise tolerance	0
init	Uniform initialization of omegas	1
tol	DC-free component	1×10^{-7}

Table 3.2 Q-learning parameter settings.

Argument	Parameter meaning	Parameter value
ε	Greedy rate	0.9
γ_m	Learning rate of rewards and punishments	0.1
γ_m	Discount factor	0.9

The LSTM model is implemented using the LSTM toolbox in MATLAB, which encapsulates the internal logic of the LSTM algorithm and exposes only the key hyperparameters for tuning. In this work, the number of hidden-layer nodes is treated as the main optimization target. The Q-learning algorithm is introduced to optimize both the VMD and LSTM model parameters, and its general hyperparameters are selected as shown in Table 3.2.

3.7.2 Algorithm Comparative Analysis

By iteratively updating the Q-table, the optimal VMD parameters are obtained as penalty factor $\eta = 3000$ and number of decomposition layers $K = 8$. These values are then used as the VMD model parameters to decompose the original manufacturing capacity sequence. The resulting modal components (IMF1–IMF8) are shown in Figure 3.2.

After decomposing the original sequence into IMF1–IMF8, each component is predicted using an LSTM model, where the number of hidden-layer nodes h is again optimized by the Q-learning algorithm. Before running Q-learning, a reasonable search range for h is specified with reference to existing literature, and the optimization is then carried out within this range.

Through this procedure, the Q-learning-guided optimization of the LSTM structure yields an optimal hidden-layer size of $h = 500$. The optimized LSTM model is used to predict each decomposed component, and the component-wise predictions are recombined to obtain the final forecast. To assess the effectiveness of the proposed Q-VMD-LSTM approach for forecasting nonstationary manufacturing capacity, two benchmark models are considered for comparison: a standard LSTM model and a VMD-LSTM model. A total of 89 data points are used for performance comparison, and the prediction outcomes obtained by all three models are shown in Figure 3.3.

In this study, three error indices were employed to quantify the prediction performance for nonstationary battery manufacturing capacity: RMSE, MAE, and MAPE. The corresponding values for each model are listed in Table 3.3.

As seen from Table 3.3, the proposed Q-VMD-LSTM algorithm achieves significantly lower RMSE, MAE, and MAPE values than the two benchmark models.

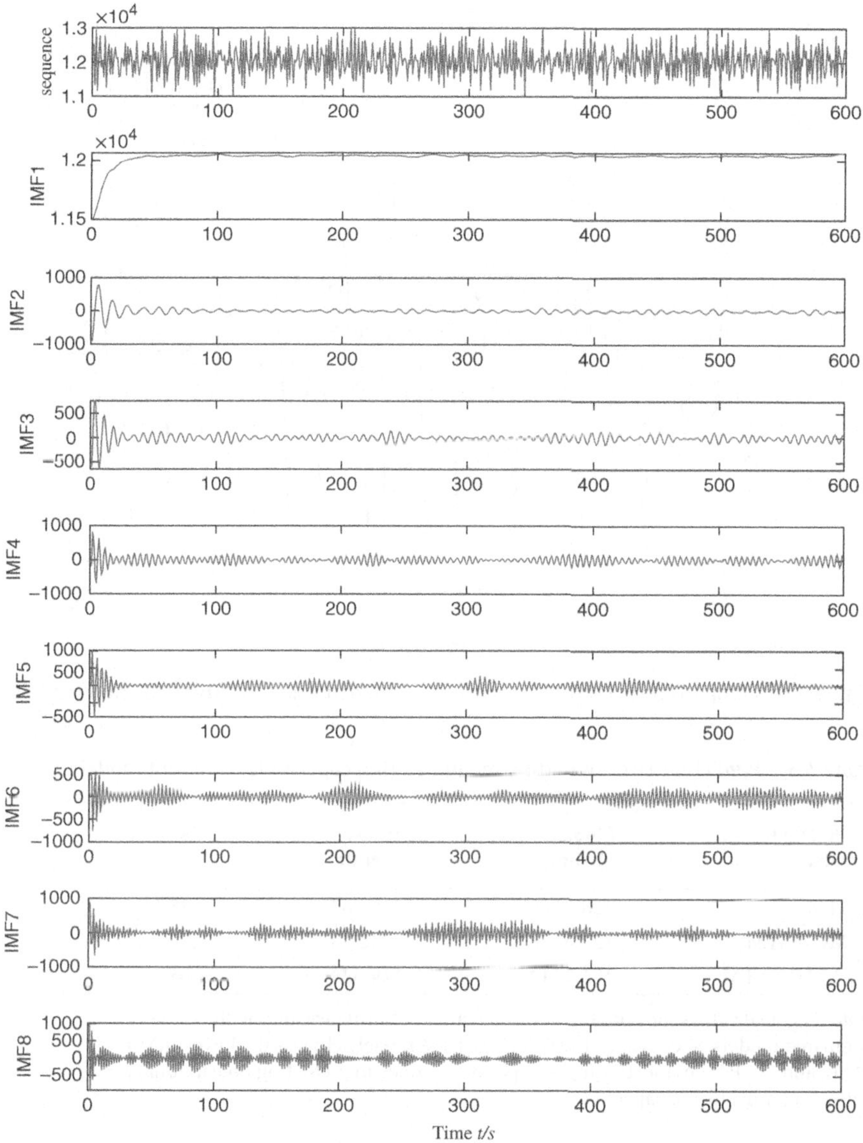

Figure 3.2 VMD decomposition result (IMF1–IMF8).

The error profiles of Q-VMD-LSTM versus VMD LSTM and Q-VMD-LSTM versus LSTM are further compared in Figures 3.4 and 3.5, respectively.

In summary, the proposed Q-VMD-LSTM method employs Q-learning to optimize the VMD parameters for modal decomposition and, subsequently, to

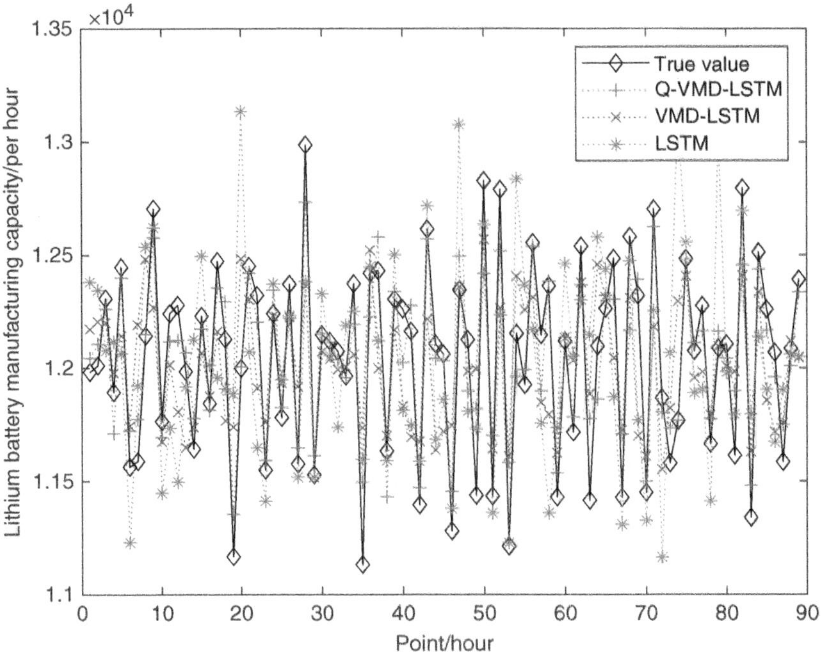

Figure 3.3 Comparison of lithium-ion battery manufacturing capacity prediction models.

Table 3.3 Prediction error of battery manufacturing capacity for different models.

Prediction model	RMSE/ quantity	MAE/ quantity	MAPE
LSTM	428.3117	345.165	2.85%
VMD-LSTM	322.4758	274.278	2.29%
Q-VMD-LSTM	**165.1213**	**129.337**	**1.08%**

Note: The bold values indicate the best prediction performance among the three compared models. Specifically, the Q-VMD-LSTM model achieves the lowest RMSE, MAE, and MAPE, demonstrating its superior accuracy in predicting nonstationary battery manufacturing capacity.

adaptively tune the LSTM prediction model, thereby forming an integrated hybrid forecasting scheme. Relative to the conventional VMD-LSTM algorithm, the RMSE is reduced from 322.4758 to 165.1213, which clearly indicates that the proposed Q-VMD-LSTM algorithm is both effective and superior to the alternatives in predicting nonstationary lithium battery manufacturing capacity.

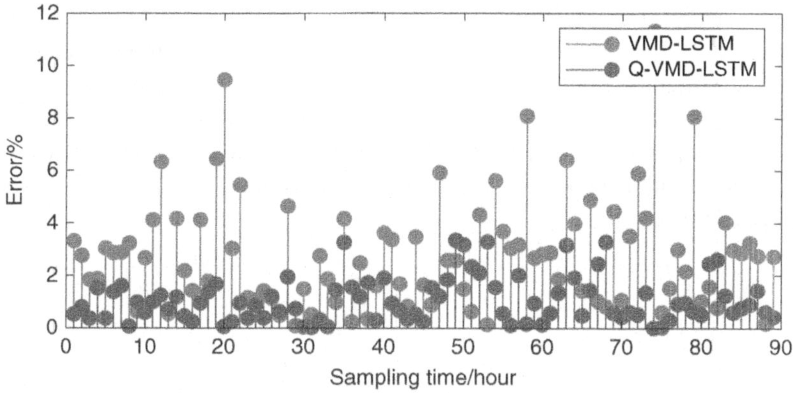

Figure 3.4 Error comparison between Q-VMD-LSTM and VMD-LSTM.

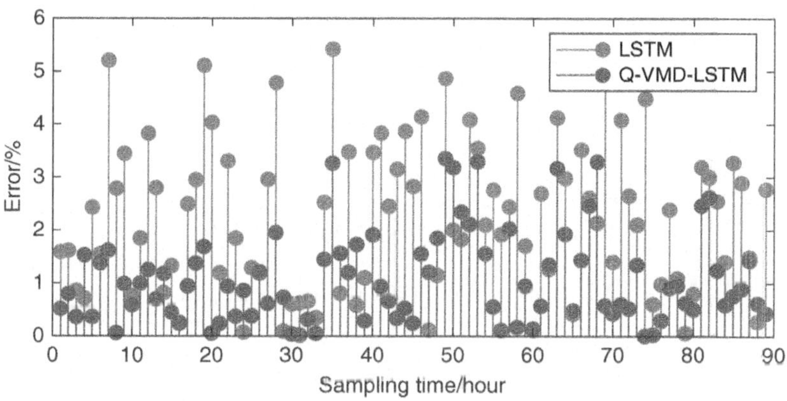

Figure 3.5 Error comparison between Q-VMD-LSTM and LSTM.

3.8 Concluding Remarks

This chapter presents the Q-VMD-LSTM hybrid prediction algorithm, which is specifically designed for forecasting nonstationary battery manufacturing capacity. The proposed methodology integrates the strengths of the Q-learning algorithm with the VMD decomposition technique, optimizing the VMD parameters adaptively to enhance the decomposition process. By leveraging the modal components extracted from the optimized VMD model, the Q-learning algorithm is then employed to fine-tune the hyperparameters of the LSTM model, which includes penalty parameters and kernel functions. This adaptive tuning significantly improves the prediction accuracy of the LSTM model for manufacturing capacity forecasting. The effectiveness of the proposed Q-VMD-LSTM framework is validated through a comprehensive comparison with several competitive algorithms, highlighting its superior performance in terms of error indices. Moreover, due to its capability to handle nonstationary and periodic data, this hybrid framework shows great potential for application in other forecasting tasks, such as wind power prediction, offering a flexible and robust solution for various time-series forecasting scenarios.

4

Advances in Battery Operating Status Analysis by Zonotopic Filtering

4.1 Introduction to Battery Operating Status Analysis

As the core energy storage unit in electric vehicles and large-scale energy storage systems, the operational state of electrochemical batteries, particularly lithium-ion batteries, directly determines overall system performance, safety, and reliability. The analysis of battery operational states constitutes a comprehensive technical framework aimed at obtaining internal key state parameters through real-time monitoring, estimation, and diagnosis. These state parameters, serving as "hidden variables" that describe the system's internal dynamics, cannot be directly measured by physical sensors, yet they are the cornerstone for efficient battery management, lifespan prediction, and proactive safety assurance.

The core of battery operational state analysis lies in the accurate online identification and tracking of the following key state variables:

- **State of Charge (SOC)**: Characterizes the remaining usable capacity of the battery. It is the direct input for energy management, range estimation, and charge–discharge strategy formulation. Accurate SOC estimation is fundamental to preventing overcharge or over-discharge, thereby ensuring the battery operates within its safe voltage window.
- **State of Health (SOH)**: Quantifies the battery's degradation level and its current maximum available capacity. It is a critical metric for assessing remaining useful life, making end-of-life decisions, and planning for second-life applications. The slow time-varying nature of SOH poses a persistent challenge to the long-term accuracy of SOC estimation.
- **State of Power (SOP)**: Defines the maximum charge and discharge power that the battery can sustainably deliver over a specific time horizon. It directly impacts vehicle dynamic performance (e.g., acceleration, regenerative braking) and the capability to provide grid services.

- **Internal States and Faults** (e.g., thermal state, incipient internal short circuit): Monitor the internal electrochemical and thermodynamic conditions of the battery. This layer is crucial for preventing catastrophic failures such as thermal runaway and enabling early safety warnings.

A battery is a nonlinear, time-varying dynamic system. Its internal characteristics are influenced by coupled factors, including operating current, ambient temperature, aging history, and C-rate, and are accompanied by significant uncertainties. These uncertainties primarily stem from two sources: (i) **stochastic uncertainties** that can be modeled as random processes but with imperfectly known statistics, and (ii) **bounded uncertainties** for which only bounds are known, not probability distributions. Traditional state-estimation methods (e.g., Kalman filter families based on Bayesian theory) rely heavily on prior assumptions about noise distributions, which are often difficult to satisfy in complex, real-world battery systems, leading to degraded accuracy or even divergence.

To overcome these limitations, set-membership estimation (SME) has gained considerable attention as a robust state-estimation paradigm. This approach only requires the assumption that system noises and disturbances are unknown but bounded (UBB). Its objective is to find a feasible set that contains all system states compatible with the model and measurements. This feasible set is typically approximated by simple geometric shapes, with mainstream methods including:

- **Interval Algorithms**: Simple in structure but often conservative in their wrapping, leading to slow convergence.
- **Ellipsoidal Algorithms**: Relatively simple to compute and iterate, sharing structural similarities with the Kalman filter. However, the inherent conservatism of ellipsoids in describing multidimensional uncertainty sets results in significant wrapping redundancy.
- **Zonotopic Algorithms**: Defined by a center and a generator matrix, zonotopes are symmetric convex polytopes that can wrap the feasible set more tightly and flexibly. They offer a superior trade-off between computational complexity and estimation accuracy, positioning them at the forefront of advanced BMS algorithm research.

This chapter adopts these principles to address the core challenges of battery systems—nonlinearity, time variance, dual uncertainties, and hard operating constraints—while naturally leading to the methodological developments in the subsequent sections. First, a baseline zonotopic filter (ZF) is formulated on the equivalent-circuit/electrothermal model under UB disturbances. Next, for systems subjected to *mixed* uncertainties (Gaussian plus UB), a zonotopic–Gaussian fusion introduces an optimal weighting to balance mean-square accuracy with set tightness. For linear parameter-varying (LPV) scenarios, an orthotope–search–expansion (OSE) strategy determines the evolution direction of time-varying parameters via measurement-strip constraints, enabling parameter–state joint estimation without the conservatism of global expansion. Finally, for nonlinear systems with explicit physical constraints (e.g., voltage and temperature bounds), a constrained zonotopic approach employs interpolation-based linearization with explicit error bounding and transforms constraints into measurement strips so that intersections with the predicted zonotope remain physically feasible.

Through comprehensive numerical simulations and experiments based on electrothermal lithium-ion models, this family of zonotopic filtering algorithms demonstrates notable advantages in estimation accuracy, computational efficiency, and robustness, outperforming traditional benchmarks such as ellipsoidal filtering.

In summary, the analysis of battery operational states forms the foundation for intelligent battery management systems (BMSs). The advanced state estimation methods based on zonotopic theory, systematically elaborated in this work, provide a rigorous and practical theoretical framework and solution suite for addressing the core challenges inherent to battery systems—nonlinearity, time-variance, dual uncertainties, and state constraints. This contribution holds substantial theoretical significance and practical value for advancing the technology of highly reliable, long-lasting BMSs.

4.2 Problem Formulation

The efficacy of a BMS is fundamentally contingent upon its ability to accurately perceive the internal key state parameters of the battery. However, the state estimation of lithium-ion batteries, which are complex electrochemical systems, faces significant challenges stemming from nonlinear dynamics, time-varying parameters, and multiple sources of uncertainty. To address these challenges systematically, this section aims to establish a unified problem formulation framework, laying the theoretical groundwork for the advanced zonotopic estimators introduced subsequently.

Battery state estimation is typically based on equivalent circuit models (ECM), which simulate the electrochemical behavior of batteries using circuit elements. The battery's ECM can be categorized into first-order and second-order models. Before discussing the second-order model, we first introduce the first-order model.

4.2.1 First-order ECM

According to the Thevenin theorem, the ECM of the battery can be obtained, and the electric model is shown in Figure 4.1. The mathematical model is expressed as:

$$\begin{cases} U = U_{oc} - U_p - I_c R_0, \\ \dot{U_p} = -\dfrac{U_p}{R_p \times C_p} + \dfrac{1}{C_p}, \end{cases} \tag{4.1}$$

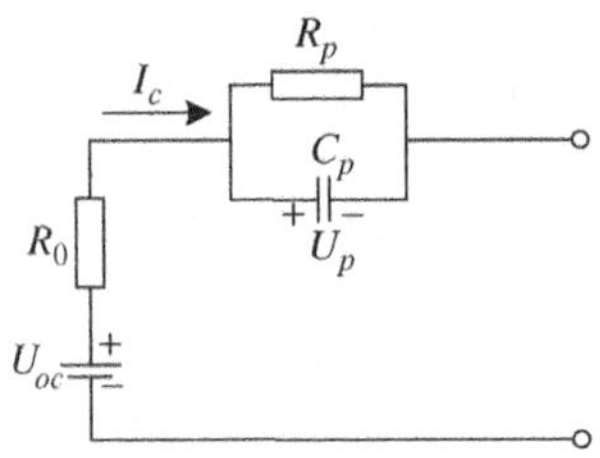

Figure 4.1 Schematic diagram of the first-order Thevenin model of Li-ion battery.

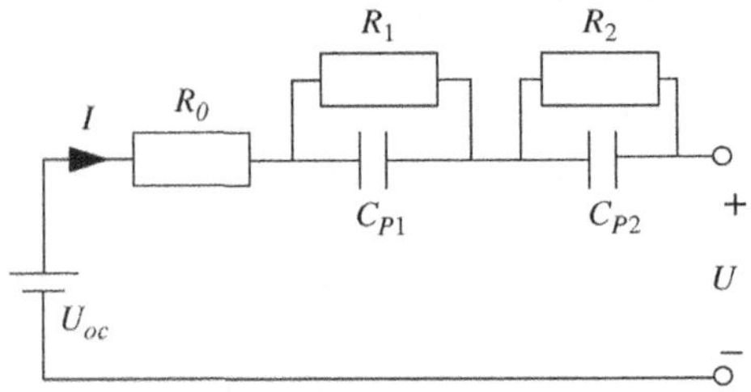

Figure 4.2 Schematic diagram of the second-order Thevenin model of Li-ion battery.

where R_p and C_p are used to describe the electrochemical polarization resistance and polarization capacitance, which vary with the battery temperature during charging and discharging, and are time-varying parameters.

4.2.2 Second-order RC ECM

Starting from the second-order RC equivalent circuit depicted in Figure 4.2, the system admits the following continuous-time dynamic equations:

$$U = U_{oc} - R_o I - U_1 - U_2,$$

$$\begin{cases} \dot{U}_1 = -\dfrac{1}{R_1 C_{p1}} U_1 + \dfrac{1}{C_{p1}} I, \\ \dot{U}_2 = -\dfrac{1}{R_2 C_{p2}} U_2 + \dfrac{1}{C_{p2}} I, \end{cases} \tag{4.2}$$

where U is the terminal voltage of the battery, U_1 and U_2 represent the electrochemical polarization voltage and concentration polarization voltage, respectively, $U_{oc}(\cdot)$ is the open-circuit voltage, typically a function of SOC, I is the battery current (positive for discharge), R_0 is the ohmic internal resistance, R_1, C_{p1}, and R_2, C_{p2} are the resistances and capacitances of the two RC networks, respectively, η is the coulombic efficiency, and Q is the battery capacity.

Discretizing the above continuous-time model (with sampling period ΔT) yields the discrete-time state-space equation:

$$\begin{bmatrix} U_{1,k} \\ U_{2,k} \\ \mathrm{SOC}_k \end{bmatrix} = \begin{bmatrix} e^{-\frac{\Delta T}{R_1 C_1}} & 0 & 0 \\ 0 & e^{-\frac{\Delta T}{R_2 C_2}} & 0 \\ 0 & 0 & 1 \end{bmatrix} \begin{bmatrix} U_{1,k-1} \\ U_{2,k-1} \\ \mathrm{SOC}_{k-1} \end{bmatrix} + \begin{bmatrix} R_1(1 - e^{-\frac{\Delta T}{R_1 C_1}}) \\ R_2(1 - e^{-\frac{\Delta T}{R_2 C_2}}) \\ -\frac{\eta \Delta T}{Q} \end{bmatrix} I_k + w_k,$$
$$U_k = U_{oc}(\mathrm{SOC}_k) - U_{1,k} - U_{2,k} - R_0 I_k + v_k, \tag{4.3}$$

where w_k stands for process noise, while v_k stands for measurement noise.

4.2.3 Battery Electrothermal Coupling Model

When considering thermal dynamics, the battery system can be modeled as an electrothermal coupling system. Using a first-order thermal model as an example, the dynamics of core temperature T_c and surface temperature T_s can be described as:

$$C_c \dot{T}_c = Q_{\text{gen}} + \frac{T_s - T_c}{R_c},$$
$$C_s \dot{T}_s = \frac{T_e - T_s}{R_u} - \frac{T_s - T_c}{R_c}, \tag{4.4}$$

where C_c and C_s represent the thermal capacities associated with the battery core and surface, respectively, while R_c is the thermal resistance from the core to the surface and R_u is the thermal resistance from the surface to the surrounding environment, T_e is the ambient temperature, and Q_{gen} is the heat generation rate, typically composed of Joule heating and reaction heat:

$$Q_{\text{gen}} = I(U_{oc} - U) = I(R_o I + U_1 + U_2). \tag{4.5}$$

Once the electrothermal coupling model is discretized, its state-space form can be written as:

$$\begin{bmatrix} T_{c,k} \\ T_{s,k} \end{bmatrix} = \begin{bmatrix} 1 - \frac{\Delta T}{R_c C_c} & \frac{\Delta T}{R_c C_c} \\ \frac{\Delta T}{R_c C_s} & 1 - \frac{\Delta T}{R_c C_s} - \frac{\Delta T}{R_u C_s} \end{bmatrix} \begin{bmatrix} T_{c,k-1} \\ T_{s,k-1} \end{bmatrix} + \begin{bmatrix} \frac{\Delta T}{C_c} & 0 \\ 0 & \frac{\Delta T}{R_u C_s} \end{bmatrix} \begin{bmatrix} Q_{\text{gen},k-1} \\ T_{e,k-1} \end{bmatrix} + w_k,$$
$$y_k = \begin{bmatrix} 1 & 0 \\ 0 & 1 \end{bmatrix} \begin{bmatrix} T_{c,k} \\ T_{s,k} \end{bmatrix} + v_k. \tag{4.6}$$

Based on the integration of the specific battery models discussed above, a unified and abstract discrete-time state-space representation can be distilled to cover various state estimation scenarios ranging from simple linear to strongly nonlinear systems. The general nonlinear form is given by:

$$\begin{cases} x_{k+1} = f(x_k) + \omega_k, \\ y_{k+1} = h(x_{k+1}) + \upsilon_{k+1}. \end{cases} \tag{4.7}$$

When the battery model is linear and explicitly includes inputs, it specializes to the common discrete linear system:

$$\begin{cases} x_{k+1} = Ax_k + Bu_k + \omega_k, \\ y_k = Cx_k + \upsilon_k, \end{cases} \tag{4.8}$$

where $x_k \in \mathbb{R}^n$ represents the system state, $u_k \in \mathbb{R}^r$ denotes the input, and $y_k \in \mathbb{R}^p$ is the measured output. The matrices A, B, and C are assumed to be known. The terms ω_k and υ_k model the process disturbance and measurement noise, respectively, satisfying the UBB condition: element-wise $|\omega_k| \leq \tilde{\omega}$ and $|\upsilon_k| \leq \tilde{\upsilon}$, where $\tilde{\omega}$ and $\tilde{\upsilon}$ are known bounds; their dimensions are compatible with the state/output equations. The functions $f(\cdot)$ and $h(\cdot)$ are discrete nonlinear mappings, reducing to the linear case in Eq. (4.8).

Concurrently, the SOH, which characterizes the battery's degradation, is equally critical. The dynamic evolution of SOH and its core coupling relationship with SOC admits the following mathematical description:

$$\begin{aligned} x_{h,k} &= x_{h,k-1} + \gamma_k, \\ d_k &= \mathrm{Soc}_k - \mathrm{Soc}_{k-1} + \frac{\eta I \Delta T}{Q_N x_{h,k-1}} + \varepsilon_k. \end{aligned} \tag{4.9}$$

Here, $x_{h,k} = \mathrm{Soh}_k \in [0, 1]$ represents the SOH, d_k is an indirect measurement related to capacity, Q_N is the nominal capacity, and η is the coulombic efficiency. Various unknown disturbances and measurement noises within the system are key factors affecting estimation accuracy and can be primarily categorized into two types: **UBB noise**, where the magnitude range is known but the distribution is unknown, such as $\omega_k \in \mathcal{W} = \langle 0, W \rangle$, $\upsilon_k \in \mathcal{V} = \langle 0, V \rangle$, $\gamma_k \in \langle 0, \Gamma \rangle$, and $\varepsilon_k \in \langle 0, \Xi \rangle$ in the equations; and **stochastic (Gaussian) noise**, with known statistical properties, often modeled as $d_k \sim \mathcal{N}(0, Q)$, $e_k \sim \mathcal{N}(0, R)$. Practical systems often exhibit a mixture of these uncertainties, placing higher demands on the robustness of the estimator.

4.3 Zonotopic Filter

4.3.1 Preliminaries

This section collects several basic notions and notations that will be used throughout this section. An interval set is written as $X = [a, b] = \{x : a \leq x \leq b\}$, where a and b denote the lower and upper bounds, respectively. The special case $\mathbf{B} = [-1, 1]$ is referred to as the unit interval, and $\mathbf{B}^m$ denotes the m-dimensional unit box obtained as the Cartesian product of m such intervals. For two sets X and Y, their Minkowski sum is defined by $X \oplus Y = \{x + y : x \in X,\ y \in Y\}$. Using this notation, a zonotope can be written in the form $\boldsymbol{p} \oplus G\mathbf{B}^m = \{\boldsymbol{p} + Gz : z \in \mathbf{B}^m\}$, where $\boldsymbol{p} \in \mathbb{R}^n$ is called the center and $G \in \mathbb{R}^{n\times m}$ is the generator matrix. The integer m is referred to as the order of the zonotope, and I_m stands for the m-dimensional identity matrix.

Several formal definitions and basic properties of zonotopes are summarized below.

Definition 4.1 [45] Given a center vector $\boldsymbol{p} \in \mathbb{R}^n$ and a collection of generator vectors $\boldsymbol{G} = \{\mathbf{g}_1, \mathbf{g}_2, \dots, \mathbf{g}_m\} \in \mathbb{R}^{n\times m}$ with $m \geq n$, the zonotope $\mathcal{Z}$ is defined as:

$$\mathcal{Z} = \boldsymbol{p} + \sum_{i=1}^{m} \alpha_i \mathbf{g}_i = \boldsymbol{p} \oplus G\mathbf{B}^m \triangleq \langle \boldsymbol{p}, G \rangle, \tag{4.10}$$

where $\alpha_i \in [-1, 1]$. In this notation, $\boldsymbol{p}$ is the center of $\mathcal{Z}$ and G is its generator matrix.

Property 1 [46] Let $\mathcal{Z}_1 = \boldsymbol{p}_1 \oplus G_1\mathbf{B}^{r_1} \subset \mathbb{R}^{n_1}$ and $\mathcal{Z}_2 = \boldsymbol{p}_2 \oplus G_2\mathbf{B}^{r_2} \subset \mathbb{R}^{n_2}$ be two zonotopes. Their Minkowski sum is again a zonotope and can be written as:

$$\mathcal{Z} = \mathcal{Z}_1 \oplus \mathcal{Z}_2 = (\boldsymbol{p}_1 + \boldsymbol{p}_2) \oplus \begin{bmatrix} G_1 & G_2 \end{bmatrix} \mathbf{B}^{r_1+r_2}. \tag{4.11}$$

Property 2 [46] Consider a zonotope $\mathcal{Z} = \boldsymbol{p} \oplus G\mathbf{B}^r \subset \mathbb{R}^n$ and a matrix K. The image of $\mathcal{Z}$ under the linear map K is again a zonotope and is given by

$$K \otimes \mathcal{Z} = (K \otimes \boldsymbol{p}) \oplus (K \otimes G)\mathbf{B}^r. \tag{4.12}$$

Lemma 4.1 *[47] For a zonotope $\mathcal{Z} = \boldsymbol{p} \oplus G\mathbf{B}^m$, the smallest axis-aligned bounding box is given by*

$$\square\mathcal{Z} = \boldsymbol{p} \oplus rs(G)\mathbf{B}^n, \tag{4.13}$$

where the diagonal matrix $rs(G)$ is defined componentwise as:

$$rs(G)_{ii} = \sum_{j=1}^{m} |G_{ij}|, \quad i = 1, \dots, n. \tag{4.14}$$

Lemma 4.2 *[48] Consider a zonotope $\mathcal{Z} = \mathbf{p} \oplus G\mathbf{B}^r \subset \mathbb{R}^n$ and an integer s with $n < s < r$. Rearrange the columns of G in nonincreasing order of their Euclidean norms and denote the resulting matrix by $\tilde{G}$. Using Lemma 2.1, one can construct an outer approximation of $\mathcal{Z}$ such that*

$$\mathcal{Z} \subseteq \mathbf{p} \oplus \begin{bmatrix} \tilde{G}_T & Q \end{bmatrix} \mathbf{B}^s,$$

where $\tilde{G}_T$ consists of the first $s - n$ columns of $\tilde{G}$, and $Q \in \mathbb{R}^{n \times n}$ is a diagonal matrix defined by

$$Q_{ii} = \sum_{j=s-n+1}^{r} |\tilde{G}_{ij}|, \quad i = 1, \dots, n. \tag{4.15}$$

Lemma 4.3 *[48] Let $\mathcal{Z} = \mathbf{p} \oplus G\mathbf{B}^r \subset \mathbb{R}^n$ be a zonotope, and consider the strip*

$$\mathcal{S} = \{x \in \mathbb{R}^n : |\mathbf{c}^{\mathrm{T}} x - d| \leq \sigma\},$$

together with a vector $\boldsymbol{\lambda} \in \mathbb{R}^n$. There exists a zonotope, parameterized by $\boldsymbol{\lambda}$, that bounds the intersection $\mathcal{Z} \cap \mathcal{S}$, namely

$$\mathcal{Z} \cap \mathcal{S} \subseteq \hat{\mathcal{Z}} = \hat{\mathbf{p}} \oplus \hat{G}\mathbf{B}^{r+1}, \tag{4.16}$$

where

$$\hat{\mathbf{p}} = \overline{\mathbf{p}} + \lambda(d - \mathbf{c}^{\mathrm{T}}\overline{\mathbf{p}}),$$
$$\hat{G} = \begin{bmatrix} (I - \lambda\mathbf{c}^{\mathrm{T}})\overline{G} & \sigma\lambda \end{bmatrix}.$$

If $\boldsymbol{\lambda}$ is chosen to minimize the size of $\hat{\mathcal{Z}}$, the optimal parameter $\boldsymbol{\lambda}^*$ is given by

$$\boldsymbol{\lambda}^* = \frac{\bar{G}\,\bar{G}^{\mathrm{T}}\mathbf{c}}{\mathbf{c}^{\mathrm{T}}\bar{G}\,\bar{G}^{\mathrm{T}}\mathbf{c} + \sigma^2}. \tag{4.17}$$

4.3.2 Zonotopic Filtering-based Operating State Estimation

The overall framework of the proposed zonotopic filtering-based operating state estimation for power batteries is illustrated in Figure 4.3. The first part of this framework is established upon the ECM of the battery. Initially, the circuit inputs are collected, and appropriate initial values for SOC and SOH are assigned. From Eqs. (4.3) and (4.9), it can be observed that SOC and SOH exhibit mutual coupling effects. Therefore, the estimation of SOC and SOH must be performed synchronously to ensure consistency and accuracy in the state estimation process. According to Eq. (4.3), the battery model can be represented as a linear discrete-time system in (4.8).

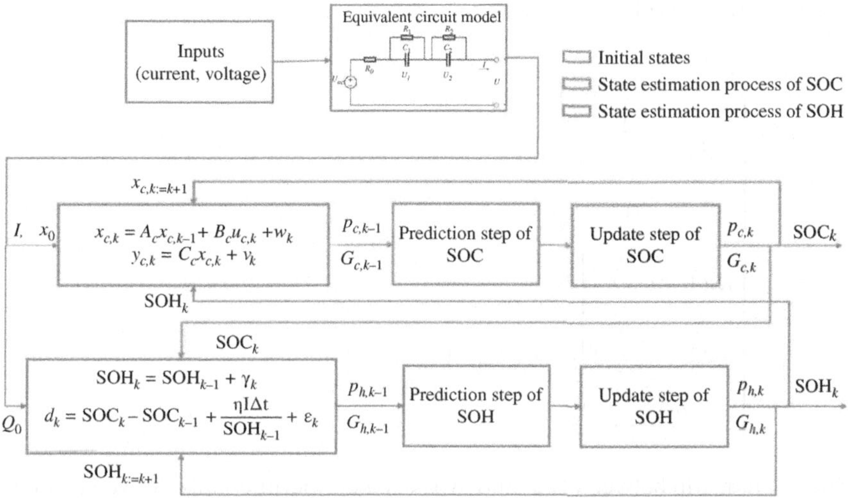

Figure 4.3 Zonotopic filtering framework for power battery state estimation at k.

4.3.2.1 Prediction Step

The prediction of the state set in zonotopic filtering can be efficiently obtained through matrix operations without approximation. The process disturbance is assumed to be bounded by a centered zonotope $\mathcal{W} = F\mathbf{B}^{n_x}$, where F denotes the generator matrix of the disturbance.

If the state at time $k-1$ is enclosed within the zonotope $\hat{\mathcal{X}}_{k-1} = \langle \hat{\mathbf{p}}_{k-1}, \hat{G}_{k-1} \rangle$, the predicted state set at time k is also a zonotope:

$$\bar{\mathcal{X}}_k = \bar{\mathbf{p}}_k \oplus \bar{G}_k \mathbf{B}^{r+n_x}, \tag{4.18}$$

where the center and generator matrix are computed as:

$$\bar{\mathbf{p}}_k = A\hat{\mathbf{p}}_{k-1} + Bu_k,$$
$$\bar{G}_k = \begin{bmatrix} A\hat{G}_{k-1} & F \end{bmatrix}.$$

For the SOC model described in Eq. (4.3), the state, input, and output vectors are defined as:

$$x_{c,k} = \begin{bmatrix} U_{1,k} \\ U_{2,k} \\ \mathrm{SOC}_k \end{bmatrix}, \quad u_k = I, \quad y_{c,k} = U_k,$$

and their corresponding matrices are

$$A_c = \begin{bmatrix} e^{-\frac{\Delta T}{\tau_1}} & 0 & 0 \\ 0 & e^{-\frac{\Delta T}{\tau_2}} & 0 \\ 0 & 0 & 1 \end{bmatrix},$$

$$B_c = \begin{bmatrix} R_1(1 - e^{-\frac{\Delta T}{\tau_1}}) \\ R_2(1 - e^{-\frac{\Delta T}{\tau_2}}) \\ \frac{\eta \Delta T}{\mathrm{SOH}_k Q_N} \end{bmatrix}, \quad C_c = \begin{bmatrix} -1 & -1 & a \end{bmatrix}.$$

The relationship between U_{OC} and SOC can be experimentally derived through multiple charge–discharge cycles, from which the OCV–SOC curve can be fitted into an approximate linear expression.

On the basis of Eq. (4.18) together with the state-space representation introduced above, the prediction step for the SOC-related zonotope yields the following center and generator matrix:

$$\begin{aligned} \overline{\boldsymbol{p}}_{c,k} &= A_c \hat{\boldsymbol{p}}_{c,k-1} + B_c u_{c,k} \\ &= \begin{bmatrix} e^{-\frac{\Delta T}{\tau_1}} & 0 & 0 \\ 0 & e^{-\frac{\Delta T}{\tau_2}} & 0 \\ 0 & 0 & 1 \end{bmatrix} x_{c,k-1} + \begin{bmatrix} R_1(1 - e^{-\frac{\Delta T}{\tau_1}}) \\ R_2(1 - e^{-\frac{\Delta T}{\tau_2}}) \\ \frac{\eta \Delta T}{\mathrm{SOH}_k Q_N} \end{bmatrix} I, \end{aligned} \tag{4.19}$$

$$\begin{aligned} \bar{G}_{c,k} &= \begin{bmatrix} A_c \hat{G}_{c,k-1} & F_c \end{bmatrix} \\ &= \begin{bmatrix} \begin{bmatrix} e^{-\frac{\Delta T}{\tau_1}} & 0 & 0 \\ 0 & e^{-\frac{\Delta T}{\tau_2}} & 0 \\ 0 & 0 & 1 \end{bmatrix} \hat{G}_{c,k-1} & F_c \end{bmatrix}, \end{aligned} \tag{4.20}$$

where the initial zonotopic center and generator for the SOC subsystem are chosen as $\hat{\boldsymbol{p}}_{c,0} = \begin{bmatrix} U_{1,0} & U_{2,0} & \mathrm{SOC}_0 \end{bmatrix}^{\mathrm{T}}$ and $\hat{G}_{c,0} = I_3 g_c$, with g_c being a prescribed constant.

For the SOH dynamics in Eq. (4.9), the state and output variables are taken as $x_{h,k} = \mathrm{SOH}_k$ and $y_{h,k} = d_k$, respectively, with the corresponding matrices $A_h = 1$ and $C_h = \frac{\eta I \Delta T}{\mathrm{SOH}_{k-1}^2}$. Accordingly, the prediction step for the SOH zonotope leads to

$$\bar{p}_{h,k} = \mathrm{SOH}_{k-1}, \tag{4.21}$$

$$\bar{G}_{h,k} = \begin{bmatrix} \hat{G}_{h,k-1} & F_h \end{bmatrix}, \tag{4.22}$$

where the initial center and generator are specified as $\hat{p}_{h,0} = \mathrm{SOH}_0$ and $\hat{G}_{h,0} = g_h$, and g_h denotes a constant parameter.

4.3.2.2 Update Step

The correction (update) step makes use of a zonotopic over-approximation of the intersection between a zonotope and a strip. The measurement noise is modeled as a symmetric interval $\mathcal{V} = \sigma \mathbf{B}^1$ with $\sigma \in \mathbb{R}^+$. Consequently, the measurement at time k can be formulated as $d = \mathbf{c}^{\mathrm{T}} x_k + v_k$, where d denotes the output and $\mathbf{c}^{\mathrm{T}}$ is the corresponding row vector.

By applying Lemma 4.3, the predicted SOC zonotope is updated through the measurement strip. The updated zonotopic center and generator for SOC are obtained as:

$$\begin{aligned} \hat{\boldsymbol{p}}_{c,k} &= \overline{\boldsymbol{p}}_{c,k} + \boldsymbol{\lambda}_c (d_c - \mathbf{c}_c^{\mathrm{T}} \overline{\boldsymbol{p}}_{c,k}) \\ &= \overline{\boldsymbol{p}}_{c,k} + \boldsymbol{\lambda}_c (y_{c,k} - C_c \overline{\boldsymbol{p}}_{c,k}) \\ &= \overline{\boldsymbol{p}}_{c,k} + \boldsymbol{\lambda}_c \left(U_k - \begin{bmatrix} -1 & -1 & a \end{bmatrix} \overline{\boldsymbol{p}}_{c,k} \right), \end{aligned} \tag{4.23}$$

$$\begin{aligned} \hat{G}_{c,k} &= \begin{bmatrix} (I_c - \boldsymbol{\lambda}_c \mathbf{c}_c^{\mathrm{T}}) \overline{G}_{c,k} & \sigma_c \boldsymbol{\lambda}_c \end{bmatrix} \\ &= \begin{bmatrix} (I_c - \boldsymbol{\lambda}_c C_c) \hat{G}_{c,k} & (y_{c,k} - C_c x_{c,k} + IR_0 - b) \boldsymbol{\lambda}_c \end{bmatrix} \\ &= \begin{bmatrix} \begin{bmatrix} 1 & 0 & 0 \\ 0 & 1 & 0 \\ 0 & 0 & 1 \end{bmatrix} - \boldsymbol{\lambda}_c \begin{bmatrix} -1 & -1 & a \end{bmatrix} \hat{G}_{c,k} \\ \left(U_k - \begin{bmatrix} -1 & -1 & a \end{bmatrix} x_{c,k} + IR_0 - b \right) \boldsymbol{\lambda}_c \end{bmatrix}^{\mathrm{T}}, \end{aligned} \tag{4.24}$$

with

$$\begin{aligned} \boldsymbol{\lambda}_c &= \frac{\bar{G}_{c,k} \bar{G}_{c,k}^{\mathrm{T}} \mathbf{c}_c}{\mathbf{c}_c^{\mathrm{T}} \bar{G}_{c,k} \bar{G}_{c,k}^{\mathrm{T}} \mathbf{c}_c + \sigma_c^2} \\ &= \frac{\bar{G}_{c,k} \bar{G}_{c,k}^{\mathrm{T}} C_c^{\mathrm{T}}}{C_c \bar{G}_{c,k} \bar{G}_{c,k}^{\mathrm{T}} C_c^{\mathrm{T}} + \sigma_c^2} \\ &= \frac{\bar{G}_{c,k} \bar{G}_{c,k}^{\mathrm{T}} \begin{bmatrix} -1 & -1 & a \end{bmatrix}^{\mathrm{T}}}{\begin{bmatrix} -1 & -1 & a \end{bmatrix} \bar{G}_{c,k} \bar{G}_{c,k}^{\mathrm{T}} \begin{bmatrix} -1 & -1 & a \end{bmatrix}^{\mathrm{T}} + \sigma_c^2}. \end{aligned} \tag{4.25}$$

Similarly, the SOH zonotope is updated as:

$$\begin{aligned}\hat{p}_{h,k} &= \overline{p}_{h,k} + \lambda_h(d_h - c_h^{\mathrm{T}}\overline{p}_{h,k}) \\ &= \overline{p}_{h,k} + \lambda_h(d_h - C_h\overline{p}_{h,k}) \\ &= \overline{p}_{h,k} + \lambda_h\left(d_k - \frac{\eta I\Delta T}{\mathrm{SOH}_{k-1}^2}\overline{p}_{h,k}\right),\end{aligned} \tag{4.26}$$

$$\begin{aligned}\hat{G}_{h,k} &= \left[(I_h - \lambda_h c_h^{\mathrm{T}})\overline{G}_{h,k} \quad \sigma_h\lambda_h\right] \\ &= \left[(I_h - \lambda_h C_h)\overline{G}_{h,k} \quad \varepsilon_k\lambda_h\right] \\ &= \begin{bmatrix} (1-\lambda_h C_h)\overline{G}_{h,k} \\ \left(d_k - \mathrm{SOC}_k + \mathrm{SOC}_{k-1} - \dfrac{\eta I\Delta T}{\mathrm{SOH}_{k-1}}\right)\lambda_h \end{bmatrix}^{\mathrm{T}},\end{aligned} \tag{4.27}$$

with

$$\begin{aligned}\lambda_h &= \frac{\bar{G}_{h,k}\bar{G}_{h,k}^{\mathrm{T}}c_h}{c_h^{\mathrm{T}}\bar{G}_{h,k}\bar{G}_{h,k}^{\mathrm{T}}c_h + \sigma_h^2} \\ &= \frac{\bar{G}_{h,k}\bar{G}_{h,k}^{\mathrm{T}}C_h^{\mathrm{T}}}{C_h\bar{G}_{h,k}\bar{G}_{h,k}^{\mathrm{T}}C_h^{\mathrm{T}} + \sigma_h^2} \\ &= \frac{\bar{G}_{h,k}\bar{G}_{h,k}^{\mathrm{T}}\left(\dfrac{\eta I\Delta T}{\mathrm{SOH}_{k-1}^2}\right)^{\mathrm{T}}}{\dfrac{\eta I\Delta T}{\mathrm{SOH}_{k-1}^2}\bar{G}_{h,k}\bar{G}_{h,k}^{\mathrm{T}}\left(\dfrac{\eta I\Delta T}{\mathrm{SOH}_{k-1}^2}\right)^{\mathrm{T}} + \sigma_h^2}.\end{aligned} \tag{4.28}$$

4.3.2.3 Operating State Estimation

Let $\mathcal{Z}_{k-1} = \langle \boldsymbol{p}_{k-1}, G_{k-1}\rangle$ be the prior zonotope; after the update one obtains $\mathcal{Z}_k = \langle \boldsymbol{p}_k, G_k\rangle$. The state vector is enclosed in the axis-aligned box $\mathrm{Box}(\mathcal{X}) = [x^-, x^+]$, whose ith upper and lower bounds are computed as:

$$\begin{cases} x^+(i) = \boldsymbol{p}(i) + \sum_{j=1}^{m}\left|G(i,j)\right|, & i = 1, \dots, n, \\ x^-(i) = \boldsymbol{p}(i) - \sum_{j=1}^{m}\left|G(i,j)\right|, & i = 1, \dots, n, \end{cases} \tag{4.29}$$

where $\boldsymbol{p}(i)$, $x^+(i)$, and $x^-(i)$ are the ith components of $\boldsymbol{p}$, x^+, and x^-, respectively, and $G(i, j)$ denotes the (i, j) entry of G.

For the SOC at time k, which corresponds to the third component of $x_{c,k}$, the resulting bounds take the form

$$\begin{cases} \mathrm{SOC}_k^+ = \overline{\boldsymbol{p}}_{c,k}(3) + \sum_{j=1}^{4k+3} \left|\overline{G}_{c,k}(3,j)\right|, \\ \mathrm{SOC}_k^- = \overline{\boldsymbol{p}}_{c,k}(3) - \sum_{j=1}^{4k+3} \left|\overline{G}_{c,k}(3,j)\right|. \end{cases} \tag{4.30}$$

In a similar manner, the SOH bounds read

$$\begin{cases} \mathrm{SOH}_k^+ = \overline{p}_{h,k} + \sum_{j=1}^{2k+1} \left|\overline{G}_{h,k}(j)\right|, \\ \mathrm{SOH}_k^- = \overline{p}_{h,k} - \sum_{j=1}^{2k+1} \left|\overline{G}_{h,k}(j)\right|. \end{cases} \tag{4.31}$$

When the generator order becomes excessive, Lemma 4.3 can be invoked to perform order reduction prior to interval extraction. The resulting upper and lower envelopes constitute robust bounds for the operating states of the power battery over time.

4.3.3 Conclusion

In this section, we develop a ZF-based state estimation framework that enables the joint estimation of both SOC and SOH of power battery systems. By employing the ECM, the method leverages ZFs to estimate the states while also calculating their upper and lower boundaries. The zonotopic filtering method offers an efficient and accurate approach, characterized by its low computational complexity, and can robustly handle measurement noise and disturbances. The proposed method addresses the challenges associated with estimating both SOC and SOH simultaneously, providing an effective solution for power battery management. Furthermore, its application extends beyond battery systems, showing promise for solving state estimation problems in other industries, such as chemical processes. The design of the ZF, along with its ability to tightly bound the state estimates, ensures reliable performance under real-world conditions. In addition to SOC and SOH estimation, the proposed method can be exploited for fault diagnosis in power battery systems [49–51] and more broadly for addressing state estimation problems arising in chemical processes.

4.4 Zonotope and Gaussian Kalman Filters

4.4.1 Preliminaries

This section introduces the mathematical preliminaries, notations, and problem setup adopted throughout the study.

First, several basic symbols are defined. The symbols $\mathbb{R}^n$ and $\mathbb{R}^{n\times m}$ denote the n-dimensional real vector space and the space of $n \times m$ real matrices, respectively. For a matrix A, A^{T} denotes its transpose and $\mathrm{tr}(A)$ its trace. For two compatible matrices A and B, the trace satisfies $\mathrm{tr}(AB) = \mathrm{tr}(B^{\mathrm{T}}A^{\mathrm{T}})$. For matrix expressions of the form AX and $AXBX^{\mathrm{T}}$, we make use of the standard identities

$$\frac{\partial \mathrm{tr}(AX)}{\partial X} = A^{\mathrm{T}}, \qquad \frac{\partial \mathrm{tr}(AXBX^{\mathrm{T}})}{\partial X} = AXB + A^{\mathrm{T}}XB^{\mathrm{T}}.$$

The operator $\mathrm{diag}(A)$ extracts the diagonal entries of A and forms a diagonal matrix. The Frobenius norm of a matrix A is defined by $\|A\|_F = \sqrt{\mathrm{tr}(A^{\mathrm{T}}A)}$.

If a scalar random variable X satisfies $X \sim \mathcal{N}(\mu, \sigma^2)$, then X is Gaussian with mean μ and variance σ^2. For two independent Gaussian random variables $X \sim \mathcal{N}(\mu_1, \sigma_1^2)$ and $Y \sim \mathcal{N}(\mu_2, \sigma_2^2)$, their sum is again Gaussian and obeys

$$X + Y \sim \mathcal{N}(\mu_1 + \mu_2, \sigma_1^2 + \sigma_2^2).$$

A zonotope $\langle p, G\rangle \subset \mathbb{R}^n$ is defined as the image of the unit hypercube $\mathbf{B}^s = [-1, +1]^s$ under an affine transformation, where $p \in \mathbb{R}^n$ is the center and $G \in \mathbb{R}^{n\times s}$ is the generator matrix. Given two zonotopes $\mathcal{Z}_1(p_1, G_1)$ and $\mathcal{Z}_2(p_2, G_2)$, their Minkowski sum is another zonotope, given by

$$\langle p_1, G_1\rangle \oplus \langle p_2, G_2\rangle = \langle p_1 + p_2, [G_1\ G_2]\rangle.$$

The linear image of a zonotope $\mathcal{Z}(p, G) \subset \mathbb{R}^n$ under a transformation matrix $M \in \mathbb{R}^{q\times n}$ is defined as:

$$M \otimes \langle p, G\rangle = \langle Mp, MG\rangle.$$

Consider the linear discrete-time system affected simultaneously by UBB disturbances and stochastic noises:

$$\begin{aligned} x_{k+1} &= Ax_k + Bu_k + w_k + d_k, \\ y_k &= Cx_k + v_k + e_k, \end{aligned} \tag{4.32}$$

where $x_k \in \mathbb{R}^n$, $u_k \in \mathbb{R}^r$, and $y_k \in \mathbb{R}^p$ denote the state, input, and measured output, respectively. The matrices A, B, and C are known and have compatible dimensions. The vectors $w_k, d_k \in \mathbb{R}^n$ represent process disturbances, whereas $v_k, e_k \in \mathbb{R}^p$ account for measurement noises.

The disturbance w_k and the measurement noise v_k are modeled as UBB signals satisfying $|w_k| \leqslant w$ and $|v_k| \leqslant v$, respectively. Equivalently, they are assumed to lie in the following zonotopic sets:

$$w_k \in \langle \mathbf{0}, W \rangle, \qquad v_k \in \langle \mathbf{0}, V \rangle,$$

with $W = \text{diag}(w)$ and $V = \text{diag}(v)$.

In addition, d_k and e_k are modeled as zero-mean Gaussian random noises,

$$d_k \sim \mathcal{N}(\mathbf{0}, Q), \qquad e_k \sim \mathcal{N}(\mathbf{0}, R),$$

where Q and R denote the covariance matrices of the stochastic process and measurement noises, respectively.

Let $\hat{x}_{k+1|k}$ and $\hat{x}_{k+1}$ denote the *a priori* and *a posteriori* state estimates, respectively. The *a priori* estimate represents the predicted state based on the process model at time k, while the *a posteriori* estimate combines the prediction and the new measurement at time $k+1$. Considering both stochastic and UBB uncertainties, these estimates can be decomposed as:

$$\hat{x}_{k+1|k} = \hat{c}_{k+1|k} + \hat{g}_{k+1|k} + \hat{z}_{k+1|k}, \tag{4.33}$$

$$\hat{x}_{k+1} = \hat{c}_{k+1} + \hat{g}_{k+1} + \hat{z}_{k+1}, \tag{4.34}$$

where $\hat{c}_{k+1|k}$ and $\hat{c}_{k+1}$ are the point estimates, $\hat{g}_{k+1|k} \sim \mathcal{N}(\mathbf{0}, P_{k+1|k})$ and $\hat{g}_{k+1} \sim \mathcal{N}(\mathbf{0}, P_{k+1})$ denote the stochastic error components, and $\hat{z}_{k+1|k} \in \mathcal{Z}(\mathbf{0}, H_{k+1|k})$, $\hat{z}_{k+1} \in \mathcal{Z}(\mathbf{0}, H_{k+1})$ represent the UBB components of the estimation error.

The initial state x_0 can be similarly decomposed as:

$$x_0 = \hat{c}_0 + \hat{g}_0 + \hat{z}_0, \tag{4.35}$$

where $\hat{c}_0$ is the nominal state, $\hat{g}_0 \sim \mathcal{N}(\mathbf{0}, P_0)$, and $\hat{z}_0 \in \mathcal{Z}(\mathbf{0}, H_0)$, with P_0 and H_0 known.

For a zonotope $\mathcal{Z}(p, G) \subset \mathbb{R}^n$, the interval hull $\text{Box}(\mathcal{Z}) = [z^-, z^+]$ gives componentwise upper and lower bounds, which are computed as:

$$\begin{cases} z^-(i) = p(i) - \sum_{j=1}^{s} |G(i,j)|, & i = 1, \dots, n, \\ z^+(i) = p(i) + \sum_{j=1}^{s} |G(i,j)|, & i = 1, \dots, n. \end{cases} \tag{4.36}$$

For a random variable $X \sim \mathcal{N}(\mu, \sigma^2)$, its confidence interval can be determined using the 3σ criterion:

$$\begin{cases} X^- = \mu - 3\sigma, \\ X^+ = \mu + 3\sigma, \end{cases} \tag{4.37}$$

where X^- and X^+ denote the lower and upper confidence bounds, respectively.

Consequently, the initial state estimate $\hat{x}_0$ admits the following lower and upper bounds:

$$\begin{cases} \hat{x}_0^- = \hat{c}_0 + \hat{g}_0^- + \hat{z}_0^-, \\ \hat{x}_0^+ = \hat{c}_0 + \hat{g}_0^+ + \hat{z}_0^+, \end{cases} \tag{4.38}$$

where $\hat{g}_0^-$ and $\hat{g}_0^+$ are obtained from the 3σ criterion, while $\hat{z}_0^-$ and $\hat{z}_0^+$ denote the lower and upper bounds associated with the zonotopic component, respectively.

The central objective of this study is to construct a hybrid state-estimation framework that integrates a polyhedron representation with Gaussian Kalman filtering. By fusing these two descriptions, the approach enables accurate and robust state estimation for linear discrete-time systems subjected simultaneously to stochastic noise and UBB disturbances.

4.4.2 Main Results

This section develops a state estimation algorithm based on the zonotope–Gaussian Kalman filter (ZGKF). The ZGKF operates in two main stages: *prediction* and *filtering*. In the prediction stage, starting from the *a posteriori* estimate at the previous step, the *a priori* state at the next step is propagated according to system (4.32), while explicitly accounting for the zonotopic bounds of the UBB disturbances, the covariance of the Gaussian noises, and the known input. In the filtering stage, the current measurement is incorporated to update the *a priori* estimate, yielding the *a posteriori* state and the associated optimal gain.

4.4.2.1 Prediction

At time $k + 1$, suppose that the *a posteriori* state estimate at time k is available and written as:

$$\hat{x}_k = \hat{c}_k + \hat{g}_k + \hat{z}_k, \tag{4.39}$$

where $\hat{g}_k \sim \mathcal{N}(\mathbf{0}, P_k)$, $\hat{z}_k \in \mathcal{Z}(\mathbf{0}, H_k)$, and the matrices P_k and H_k are known.

By the state transition in (4.32), the *a priori* estimate at time $k + 1$ is given by

$$\hat{x}_{k+1|k} = A\hat{x}_k + Bu_k + w_k + d_k. \tag{4.40}$$

The decomposition of $\hat{x}_{k+1|k}$ is characterized by the following theorem.

Theorem 4.1 *Consider* (4.32). *Suppose $\hat{x}_k$ is known at time k. Then the a priori estimate at time $k + 1$ admits the decomposition*

$$\hat{x}_{k+1|k} = \hat{c}_{k+1|k} + \hat{g}_{k+1|k} + \hat{z}_{k+1|k}, \tag{4.41}$$

where

$$\hat{c}_{k+1|k} = A\hat{c}_k + Bu_k, \tag{4.42}$$
$$\hat{g}_{k+1|k} \sim \mathcal{N}(\mathbf{0}, P_{k+1|k}), \tag{4.43}$$
$$\hat{z}_{k+1|k} \in \mathcal{Z}(\mathbf{0}, H_{k+1|k}), \tag{4.44}$$
$$P_{k+1|k} = AP_kA^{\mathrm{T}} + Q, \tag{4.45}$$
$$H_{k+1|k} = [AH_k \quad W]. \tag{4.46}$$

Proof 4.1. *Given $\hat{x}_k$, we have*

$$\begin{aligned} \hat{x}_{k+1|k} &= A\hat{x}_k + Bu_k + w_k + d_k \\ &= A(\hat{c}_k + \hat{g}_k + \hat{z}_k) + Bu_k + w_k + d_k \\ &= \hat{c}_{k+1|k} + \hat{g}_{k+1|k} + \hat{z}_{k+1|k}, \end{aligned} \tag{4.47}$$

with

$$\hat{c}_{k+1|k} = A\hat{c}_k + Bu_k, \tag{4.48}$$
$$\hat{g}_{k+1|k} = A\hat{g}_k + d_k, \tag{4.49}$$
$$\hat{z}_{k+1|k} = A\hat{z}_k + w_k. \tag{4.50}$$

Since $\hat{g}_k \sim \mathcal{N}(\mathbf{0}, P_k)$ and $d_k \sim \mathcal{N}(\mathbf{0}, Q)$, it follows that

$$\hat{g}_{k+1|k} \sim \mathcal{N}(\mathbf{0}, P_{k+1|k}), \tag{4.51}$$
$$P_{k+1|k} = AP_kA^{\mathrm{T}} + Q. \tag{4.52}$$

Moreover, from $\hat{z}_k \in \mathcal{Z}(\mathbf{0}, H_k)$ and $w_k \in \mathcal{Z}(\mathbf{0}, W)$ and the closure of zonotopes under linear maps and Minkowski sums,

$$\begin{aligned} \hat{z}_{k+1|k} &= A\hat{z}_k + w_k \\ &\in \langle \mathbf{0}, AH_k \rangle \oplus \langle \mathbf{0}, W \rangle \\ &= \mathcal{Z}\langle \mathbf{0}, H_{k+1|k} \rangle, \end{aligned} \tag{4.53}$$

where $H_{k+1|k} = [AH_k \ \ W]$.

4.4.2.2 Filtering

Having obtained $\hat{x}_{k+1|k}$, we next fuse the new measurement y_{k+1} to form the *a posteriori* estimate $\hat{x}_{k+1}$.

Theorem 4.2 *The* a posteriori *estimate admits the form* $\hat{x}_{k+1} = \hat{c}_{k+1} + \hat{g}_{k+1} + \hat{z}_{k+1}$, *where*

$$\hat{c}_{k+1} = (I - K_{k+1}C)\hat{c}_{k+1|k} + K_{k+1}y_{k+1}, \tag{4.54}$$

$$\hat{g}_{k+1} \sim \mathcal{N}(\mathbf{0}, P_{k+1}), \tag{4.55}$$

$$\hat{z}_{k+1} \in \mathcal{Z}(\mathbf{0}, H_{k+1}), \tag{4.56}$$

$$P_{k+1} = (I - K_{k+1}C)P_{k+1|k}(I - K_{k+1}C)^{\mathrm{T}} + K_{k+1}RK_{k+1}^{\mathrm{T}}, \tag{4.57}$$

$$H_{k+1} = \left[(I - K_{k+1}C)H_{k+1|k} \quad K_{k+1}V\right]. \tag{4.58}$$

Proof 4.2. By the standard innovation update,

$$\begin{aligned}
\hat{x}_{k+1} &= \hat{x}_{k+1|k} + K_{k+1}(y_{k+1} - C\hat{x}_{k+1|k}) \\
&= (I - K_{k+1}C)\hat{x}_{k+1|k} + K_{k+1}y_{k+1} \\
&= (I - K_{k+1}C)(\hat{c}_{k+1|k} + \hat{g}_{k+1|k} + \hat{z}_{k+1|k}) \\
&\quad + K_{k+1}(Cx_{k+1} + v_{k+1} + e_{k+1}) \\
&= \hat{c}_{k+1} + \hat{g}_{k+1} + \hat{z}_{k+1},
\end{aligned} \tag{4.59}$$

where K_{k+1} *is the gain and*

$$\hat{c}_{k+1} = (I - K_{k+1}C)\hat{x}_{k+1|k} + K_{k+1}Cx_{k+1}, \tag{4.60}$$

$$\hat{g}_{k+1} = (I - K_{k+1}C)\hat{g}_{k+1|k} + K_{k+1}e_{k+1}, \tag{4.61}$$

$$\hat{z}_{k+1} = (I - K_{k+1}C)\hat{z}_{k+1|k} + K_{k+1}v_{k+1}. \tag{4.62}$$

Since $\hat{c}_{k+1}$ *represents the point estimate,* v_{k+1} *and* e_{k+1} *are set to zero therein, yielding*

$$\hat{c}_{k+1} = (I - K_{k+1}C)\hat{c}_{k+1|k} + K_{k+1}y_{k+1}. \tag{4.63}$$

With $\hat{g}_{k+1|k} \sim \mathcal{N}(\mathbf{0}, P_{k+1|k})$ *and* $e_{k+1} \sim \mathcal{N}(\mathbf{0}, R)$, *the stochastic part satisfies*

$$\hat{g}_{k+1} \sim \mathcal{N}(\mathbf{0}, P_{k+1}), \tag{4.64}$$

$$P_{k+1} = (I - K_{k+1}C)P_{k+1|k}(I - K_{k+1}C)^{\mathrm{T}} + K_{k+1}RK_{k+1}^{\mathrm{T}}. \tag{4.65}$$

Finally, using $\hat{z}_{k+1|k} \in \mathcal{Z}(\mathbf{0}, H_{k+1|k})$, $v_{k+1} \in \mathcal{Z}(\mathbf{0}, V)$ *and zonotope closure,*

$$\begin{aligned}
\hat{z}_{k+1} &= (I - K_{k+1}C)\hat{z}_{k+1|k} + K_{k+1}v_{k+1} \\
&\in \langle \mathbf{0}, (I - K_{k+1}C)H_{k+1|k}\rangle \oplus \langle \mathbf{0}, K_{k+1}V\rangle \\
&= \mathcal{Z}\langle \mathbf{0}, H_{k+1}\rangle,
\end{aligned} \tag{4.66}$$

with $H_{k+1} = \left[(I - K_{k+1}C)H_{k+1|k} \;\; K_{k+1}V\right]$.

To improve the accuracy of the state estimate, it is essential to design an appropriate gain factor K_{k+1}. Analogous to the standard KF derivation, we compute K_{k+1} by minimizing the mean-square error (MSE) of the *a posteriori* estimate. For system (4.32), the MSE consists of a stochastic term and a UBB-uncertainty term:

$$\mathrm{E}\left[\|\hat{x}_{k+1} - x_{k+1}\|^2\right] = \mathrm{E}\left[\left\|\Delta_{k+1}^{\mathrm{stoc}}\right\|^2\right] + \left\|\Delta_{k+1}^{\mathrm{UBB}}\right\|^2. \tag{4.67}$$

For the stochastic term $\mathrm{E}\left[\left\|\Delta_{k+1}^{\mathrm{stoc}}\right\|^2\right]$, the MSE is captured by the trace of the covariance matrix P_{k+1}. For the UBB component $\left\|\Delta_{k+1}^{\mathrm{UBB}}\right\|^2$, an upper bound on the squared error is provided by the Frobenius norm of the generator matrix H_{k+1}. Hence, $\mathrm{E}\left[\left\|\Delta_{k+1}^{\mathrm{stoc}}\right\|^2\right] = \mathrm{tr}(P_{k+1})$ and $\left\|\Delta_{k+1}^{\mathrm{UBB}}\right\|^2 = \|H_{k+1}\|_F^2 = \mathrm{tr}\left(H_{k+1}^{\mathrm{T}} H_{k+1}\right)$.

Define the composite objective

$$J = (1 - \eta_{k+1})J_g + \eta_{k+1}J_z, \tag{4.68}$$

where $J_g = \mathrm{tr}(P_{k+1})$ and $J_z = \mathrm{tr}\left(H_{k+1}^{\mathrm{T}} H_{k+1}\right)$. The weight $\eta_{k+1} \in [0, 1]$ balances the relative emphasis on J_g and J_z and thereby determines the gain K_{k+1}. For instance, $\eta_{k+1} = \frac{1}{2}$ assigns equal weight; $\eta_{k+1} = 1$ corresponds to a purely UBB setting; $\eta_{k+1} = 0$ recovers the purely stochastic case.

Theorem 4.3 *Given P_{k+1} in (4.57), H_{k+1} in (4.58), and J in (4.68), the optimal gain K_{k+1} is*

$$\begin{aligned} K_{k+1} = {} & \left[(1 - \eta_{k+1})P_{k+1|k}C^{\mathrm{T}} - \eta_{k+1}M_{k+1}a_{k+1}\right]\left[(1 - \eta_{k+1})R \right. \\ & \left. + (1 - \eta_{k+1})CP_{k+1|k}C^{\mathrm{T}} + \eta_{k+1}a_{k+1}^{\mathrm{T}}a_{k+1}\right]^{-1}, \end{aligned} \tag{4.69}$$

where $M_{k+1} = [H_{k+1|k} \quad \mathbf{0}]$, $a_{k+1}^{\mathrm{T}} = [-CH_{k+1|k} \quad V]$, and

$$\eta_{k+1} = \frac{\mathrm{tr}\left(H_k^{\mathrm{T}} H_k\right)}{\mathrm{tr}(P_k) + \mathrm{tr}\left(H_k^{\mathrm{T}} H_k\right)}. \tag{4.70}$$

Proof 4.3. Minimizing J yields

$$\frac{\partial J}{\partial K_{k+1}} = (1 - \eta_{k+1})\frac{\partial J_g}{\partial K_{k+1}} + \eta_{k+1}\frac{\partial J_z}{\partial K_{k+1}} = 0. \tag{4.71}$$

Substituting (4.57) into J_g:

$$\begin{aligned} J_g &= \mathrm{tr}\left[(I - K_{k+1}C)P_{k+1|k}(I - K_{k+1}C)^{\mathrm{T}} + K_{k+1}RK_{k+1}^{\mathrm{T}}\right] \\ &= \mathrm{tr}(P_{k+1|k} - P_{k+1|k}C^{\mathrm{T}}K_{k+1}^{\mathrm{T}} - K_{k+1}CP_{k+1|k} \\ &\quad + K_{k+1}CP_{k+1|k}C^{\mathrm{T}}K_{k+1}^{\mathrm{T}} + K_{k+1}RK_{k+1}^{\mathrm{T}}), \end{aligned} \tag{4.72}$$

and the trace-derivative identities give

$$\frac{\partial J_g}{\partial K_{k+1}} = 2K_{k+1}CP_{k+1|k}C^{\mathrm{T}} + 2K_{k+1}R - 2P_{k+1|k}C^{\mathrm{T}}. \tag{4.73}$$

Similarly, with the decomposition

$$H_{k+1} = M_{k+1} + K_{k+1}a_{k+1}^{\mathrm{T}}, \tag{4.74}$$

where $M_{k+1} = [H_{k+1|k} \quad \mathbf{0}]$ *and* $a_{k+1}^{\mathrm{T}} = [-CH_{k+1|k} \quad V]$, *we have*

$$\begin{aligned}
&\|H_{k+1}\|_F^2 = \left\|M_{k+1} + K_{k+1}a_{k+1}^{\mathrm{T}}\right\|_F^2 \\
&= \mathrm{tr}\left[\left(M_{k+1}^{T} + a_{k+1}K_{k+1}^{\mathrm{T}}\right)\left(M_{k+1} + K_{k+1}a_{k+1}^{\mathrm{T}}\right)\right] \\
&= \mathrm{tr}\left(M_{k+1}^{\mathrm{T}}M_{k+1}\right) + \mathrm{tr}\left(a_{k+1}K_{k+1}^{\mathrm{T}}M_{k+1}\right) \\
&\quad + \mathrm{tr}\left(M_{k+1}^{\mathrm{T}}K_{k+1}a_{k+1}^{\mathrm{T}}\right) + \mathrm{tr}\left(a_{k+1}K_{k+1}^{\mathrm{T}}K_{k+1}a_{k+1}^{\mathrm{T}}\right) \\
&= 2K_{k+1}^{\mathrm{T}}M_{k+1}a_{k+1} + a_{k+1}^{\mathrm{T}}a_{k+1}K_{k+1}^{\mathrm{T}}K_{k+1} + \mathrm{tr}\left(M_{k+1}^{\mathrm{T}}M_{k+1}\right).
\end{aligned} \tag{4.75}$$

so that

$$\frac{\partial J_z}{\partial K_{k+1}} = 2M_{k+1}a_{k+1} + 2K_{k+1}a_{k+1}^{\mathrm{T}}a_{k+1}. \tag{4.76}$$

Combining (4.73) and (4.76) and setting $\partial J/\partial K_{k+1} = 0$ *gives*

$$\begin{aligned}
&\eta_{k+1}\left(2M_{k+1}a_{k+1} + 2K_{k+1}a_{k+1}^{\mathrm{T}}a_{k+1}\right) \\
&\quad + (1-\eta_{k+1})(2K_{k+1}CP_{k+1|k}C^{\mathrm{T}} + 2K_{k+1}R - 2P_{k+1|k}C^{\mathrm{T}}) = 0,
\end{aligned} \tag{4.77}$$

from which (4.69) follows directly. The adaptive weight (4.70) balances the stochastic and UBB contributions.

Finally, the interval bounds of $\hat{x}_{k+1}$ are obtained as:

$$\begin{cases}
\hat{x}_{k+1}^{-} = \hat{c}_{k+1} + \hat{g}_{k+1}^{-} + \hat{z}_{k+1}^{-}, \\
\hat{x}_{k+1}^{+} = \hat{c}_{k+1} + \hat{g}_{k+1}^{+} + \hat{z}_{k+1}^{+}.
\end{cases} \tag{4.78}$$

For system (4.32), the complete ZGKF-based state estimation procedure is outlined in Algorithm 1.

Algorithm 1: ZGKF algorithm

Require: Initial values: $c_0^+, P_0, H_0, W, V, Q, R$

1: $N \leftarrow$ Constant1

2: **for** $k = 1 : N$ **do**

3: Acquire input u_{k-1} and measurement $y(k)$

4: Propagate the *a priori* quantities:

$$\begin{aligned} \hat{c}_{k|k-1} &= A\hat{c}_{k-1} + Bu_{k-1} \\ P_{k|k-1} &= AP_{k-1}A^{\mathrm{T}} + Q \\ H_{k|k-1} &= [AH_{k-1} \quad W] \end{aligned}$$

5: Update the weight

$$\eta_k = \frac{\mathrm{tr}\left(H_{k-1}^{\mathrm{T}} H_{k-1}\right)}{\mathrm{tr}(P_{k-1}) + \mathrm{tr}\left(H_{k-1}^{\mathrm{T}} H_{k-1}\right)}$$

6: Determine the optimal gain K_k using (4.69)

7: Form the *a posteriori* quantities:

$$\begin{aligned} \hat{c}_k &= (I - K_k C)\hat{c}_{k|k-1} + K_k C x_k \\ P_k &= (I - K_k C)P_{k|k-1} + K_k R K_k^{\mathrm{T}} \\ H_k &= \left[(I - K_k C)H_{k|k-1} \quad K_k V\right] \end{aligned}$$

8: Compute the interval of $\hat{x}_k$:

$$\begin{cases} \hat{x}_k^- = \hat{c}_k + \hat{g}_k^- + \hat{z}_k^-, \\ \hat{x}_k^+ = \hat{c}_k + \hat{g}_k^+ + \hat{z}_k^+ \end{cases}$$

9: **return** $\hat{c}_k, P_k, H_k$

10: **end for**

4.4.3 Conclusion

In this section, a hybrid state estimation scheme has been developed in which a zonotopic description of UBB disturbances is combined with a Gaussian Kalman filter for stochastic noises. The resulting ZGKF operates by propagating, at each step, both a stochastic covariance and a polyhedral (zonotopic) uncertainty set, and by designing a gain that simultaneously accounts for the two noise types. In particular, the gain is obtained by minimizing a performance index that includes the mean-square error term associated with the covariance matrix and a term measuring the size of the zonotopic set, so that the estimator balances accuracy in the probabilistic sense with tightness of the feasible region.

The proposed ZGKF framework yields a tighter and more informative characterization of the system state than classical purely stochastic filters, because it exploits prior bounds on UBB disturbances while retaining the advantages of Gaussian Kalman filtering. At the same time, the algorithm preserves a moderate computational burden, making it suitable for real-time implementation in linear discrete-time systems subject to mixed stochastic and UBB uncertainties. Although the method is illustrated on the battery-related example in this chapter, the underlying formulation is generic and can be transferred to other applications that require robust state estimation under dual uncertainty, such as radar tracking, integrated navigation, and related fields [52].

4.5 OSE-based Extended Zonotopic Kalman Filter

4.5.1 Preliminaries and Problem Description

4.5.1.1 Preliminaries

Definition 4.2 The symbol $\oplus$ denotes the Minkowski sum. The matrix $I_n \in \mathbb{R}^{n\times n}$ denotes the identity of order n.

Definition 4.3 Let $\mathcal{Z} = p \oplus G\mathbf{B}^s \subset \mathbb{R}^n$ be an s-order zonotope. A reduction operator produces a lower-dimensional outer approximation

$$\mathcal{Z} = \langle p, G\rangle \subseteq \langle p, G_{\downarrow_q}\rangle, \tag{4.79}$$

where $G_{\downarrow_q}$ limits the number of columns to at most q to reduce complexity.

Sorting the columns of G in nonincreasing order of their Euclidean norms produces the matrix $\tilde{G}$. For any integer q satisfying $n \leqslant q \leqslant s$, a reduced generator matrix $G_{\downarrow_q}$ is constructed as $G_{\downarrow_q} = [\, \tilde{G}_T,\ Q\,]\mathbf{B}^s$, where $\tilde{G}_T$ contains the first $s-n$ columns of $\tilde{G}$ and Q is a diagonal matrix defined in [48] by

$$Q(i,i) = \sum_{j=1}^{s-n+1} \left|\tilde{G}(i,j)\right|, \quad i = 1, \dots, n. \tag{4.80}$$

When $q = 1$, the reduced generator $G_{\downarrow_1}$ degenerates to a single column and satisfies

$$p \oplus G\mathbf{B}^s \subseteq p \oplus G_{\downarrow_1}\mathbf{B}^1, \tag{4.81}$$

where $G_{\downarrow_1}\mathbf{B}^1$ defines a line segment, and its ith component is given by

$$G_{\downarrow_1}(i) = \sum_{j=1}^{s} \left|G(i,j)\right|, \quad i = 1, \dots, n. \tag{4.82}$$

Definition 4.4 The zonotope $\mathcal{Z} = p \oplus G\mathbf{B}^s \subset \mathbb{R}^n$ admits a smallest axis-aligned bounding box [50]:

$$\mathcal{Z} \subseteq \square\mathcal{Z} = [\, \underline{z},\ \overline{z}\,] = \langle p, G_{\downarrow n} \rangle, \tag{4.83}$$

where $\overline{z}$ and $\underline{z}$ stand for the upper and lower endpoints of the interval box $\square\mathcal{Z}$, and the diagonal matrix $G_{\downarrow n}$ is specified by

$$G_{\downarrow n}(i, i) = \sum_{j=1}^{s} \left| G(i, j) \right|, \quad i = 1, \ldots, n. \tag{4.84}$$

Definition 4.5 The Frobenius radius is used to characterize the spatial size of a zonotope $\mathcal{Z} = p \oplus G\mathbf{B}^s \subset \mathbb{R}^n$ and is defined in [53] as:

$$\|\langle p, G \rangle\|_F = \|G\|_F = \sqrt{\mathrm{tr}(G^{\mathrm{T}}G)} = \sqrt{\mathrm{tr}(GG^{\mathrm{T}})}. \tag{4.85}$$

A weighted Frobenius radius can be introduced as $\|G\|_{F,M} = \sqrt{\mathrm{tr}(G^{\mathrm{T}}MG)}$, where $M \in \mathbb{R}^{n\times n}$ is a symmetric positive definite weighting matrix.

Definition 4.6 For a vector $p \in \mathbb{R}^n$ or a matrix $G \in \mathbb{R}^{n\times s}$, the operator $[\cdot]^+$ denotes the componentwise absolute value [54], that is,

$$[p]^+(i) = |p(i)|, \quad i = 1, \ldots, n, \tag{4.86}$$

$$[G]^+(i, j) = |G(i, j)|, \quad j = 1, \ldots, s. \tag{4.87}$$

Property 3 For compatible matrices A, B, X, the following identities hold [55]:

$$\mathrm{tr}(A) = \mathrm{tr}(A^{\mathrm{T}}), \tag{4.88}$$

$$\mathrm{tr}(AB) = \mathrm{tr}(B^{\mathrm{T}}A^{\mathrm{T}}), \tag{4.89}$$

$$\frac{\partial\, \mathrm{tr}(AX)}{\partial X} = A^{\mathrm{T}}, \tag{4.90}$$

$$\frac{\partial\, \mathrm{tr}(AXBX^{\mathrm{T}})}{\partial X} = AXB + A^{\mathrm{T}}XB^{\mathrm{T}}. \tag{4.91}$$

Definition 4.7 For intervals $[a, b]$ and $[c, d]$, the arithmetic operations are defined as [56]:

$$[a, b] + [c, d] = [a + c,\ b + d], \tag{4.92}$$

$$[a, b] - [c, d] = [a - d,\ b - c], \tag{4.93}$$

$$[a, b] \times [c, d] = [\min(ac, ad, bc, bd),\ \ \max(ac, ad, bc, bd)], \tag{4.94}$$

$$\frac{[a, b]}{[c, d]} = [a, b] \times \left[\frac{1}{d}, \frac{1}{c}\right], \qquad 0 \notin [c, d]. \tag{4.95}$$

4.5.1.2 Problem Description

Owing to the main electrical parameters of the battery model (such as R_0, R_1, C_1) are not constant but vary significantly with operating conditions, temperature and aging, a fixed-parameter model is no longer sufficient. To capture these variations, we adopt an LPV description with dual noise terms:

$$\begin{cases} x(k+1) = A(\theta(k))x(k) + B(\theta(k))u(k) + Ew(k), \\ \quad y(k) = Cx(k) + Fv(k), \end{cases} \tag{4.96}$$

where $x(k) \in \mathbb{R}^{n_x}$ and $y(k) \in \mathbb{R}^{n_y}$ denote the state vector and the measured output, respectively, and $u(k) \in \mathbb{R}^{u}$ is the known input of the LPV system. The matrices E and F are known. The signals $w(k)$ and $v(k)$ satisfy the UBB conditions $|w(k)|_\infty \leqslant w \in \mathbb{R}^{n_x}$ and $|v(k)|_\infty \leqslant v \in \mathbb{R}^{n_y}$. The matrices $A(\theta(k))$ and $B(\theta(k))$ depend on a time-varying parameter vector $\theta(k)$; their dimensions are fixed, but their entries are functions of $\theta(k)$ and are assumed to be identifiable.

The evolution of the parameter vector and its measurement model are described by

$$\begin{cases} \theta(k) = \theta(k-1) + \delta(k), \\ h(k) = \varphi^{\mathrm{T}}(k)\theta(k) + e(k), \end{cases} \tag{4.97}$$

where $\theta(k) \in \mathbb{R}^{n_\theta}$ collects the unknown parameters, $\varphi(k) \in \mathbb{R}^{n_\theta}$ is the regression vector, $e(k)$ denotes the identification noise with $|e(k)|_\infty \leqslant \sigma$, and $\delta(k)$ describes the variation of the time-varying parameters with $|\delta(k)|_\infty \leqslant \gamma$.

Since the identification noise is bounded, the admissible parameter values at time k are confined to a strip,

$$S(k) = \{\theta(k) : |h(k) - \varphi^{\mathrm{T}}(k)\theta(k)| \leqslant \sigma\}, \tag{4.98}$$

which can be interpreted as the region between two parallel hyperplanes in the parameter space:

$$\begin{aligned} S_1(k) &= \{\theta(k) : h(k) = \varphi^{\mathrm{T}}(k)\theta(k) - \sigma\}, \\ S_2(k) &= \{\theta(k) : h(k) = \varphi^{\mathrm{T}}(k)\theta(k) + \sigma\}. \end{aligned} \tag{4.99}$$

The feasible parameter set at time k is then described by the polyhedral set

$$\Theta(k) = \{\theta \in \mathbb{R}^{n_\theta} : \alpha_k\theta \leqslant \beta_k, \qquad k = 1, \dots, N\}, \tag{4.100}$$

where $\alpha_k \in \mathbb{R}^{n \times n_\theta}$ and $\beta_k \in \mathbb{R}^{n}$ specify the linear inequality constraints. To obtain the smallest orthotope $O(k)$ that contains $\Theta(k)$, one has to solve $2n_\theta$ linear programming problems. The admissible range of the parameter variation is determined by the bound $\gamma \in \mathbb{R}^{+}$.

From (4.96), the initial state x_0 and the noise sequences w_k and v_k are UBB. Using the zonotope representation, these quantities are enclosed in the following sets:

$$x(0) \in \mathcal{Z}_0 = \langle p(0), B(0) \rangle, \quad w(k) \in W = \langle 0, G_w \rangle, \quad v(k) \in V = \langle 0, G_v \rangle, \tag{4.101}$$

where $\mathcal{Z}_0$ is the initial state zonotope; $W = \mathrm{diag}(w)$ and $V = \mathrm{diag}(v)$ are known diagonal matrices, and G_w, G_v are the corresponding generator matrices defining the zonotopes W and V.

The observer-based estimator associated with system (4.96) is chosen as:

$$\hat{x}(k+1) = A(\theta(k))\hat{x}(k) + B(\theta(k))u(k) + L\left[y(k) - C\hat{x}(k) - Fv(k)\right], \tag{4.102}$$

where $\hat{x}(k) \in \mathbb{R}^{n_x}$ denotes the state estimate and L is the estimator gain to be designed.

The aim of this section is to construct a state-estimation scheme for the discrete-time LPV system (4.96) based on set-membership principles. The proposed approach relies on designing an optimal orthotopic region for the parameter vector together with an extended zonotopic Kalman filter (EZKF). The uncertain parameters are confined within orthotopic linear programs, and all uncertain states are enclosed by optimally tightened zonotopes, thereby reducing the impact of parameter variability on the estimation performance.

4.5.2 Main Results

In this section, we investigate the identification of uncertain parameters in discrete time LPV systems and develop an orthotopic spatial search expansion algorithm. Building on the orthotopic sets generated by this procedure, an optimal EZKF is subsequently constructed.

4.5.2.1 OSE-algorithm-based Time-varying Parameter Identification

Lemma 4.4 *Computing the minimum orthotope O_k requires solving $2n_\theta$ linear constraint equations via linear programming. A generic linear constraint problem is formulated as [57]:*

$$\begin{aligned} &\max \quad c^{\mathrm{T}}\theta, \\ &\quad \text{s.t.} \\ &\quad \alpha\theta \leqslant \beta, \\ &\theta \in \mathcal{O} = \{\theta : a_i^{\mathrm{T}}\theta \leqslant \beta,\ i \in \mathcal{I}\}, \end{aligned}$$

where a_i^{T} denotes the ith row of the matrix α, and $\mathcal{O}$ is the corresponding bounding set.

Starting from the orthotopic description in (4.100), an enclosing orthotope O_k for the feasible parameter set is obtained by solving $2n_\theta$ linear programs. The orthotope is written as:

$$\theta(k) \in O(\bar{\theta}(k), d(k)) = \{\theta(k) : \theta(k) = \bar{\theta}(k) + \mathrm{diag}(d(k))\omega, \ \|\omega\|_\infty \leqslant 1\}, \tag{4.103}$$

where $\bar{\theta}(k) \in \mathbb{R}^{n_\theta}$ is the center, $d(k) \in \mathbb{R}^{n_\theta}$ collects the half-widths, $\omega \in \mathbb{R}^{n_\theta}$ is an auxiliary vector, and $\mathrm{diag}(d(k))$ is the diagonal matrix whose diagonal entries are $d_i(k)$, $i = 1, \dots, n_\theta$. The $2n_\theta$ linear optimization problems are formulated as:

$$\begin{aligned} \overline{d}_i(k) = \arg\max \quad & c_i^{\mathrm{T}}\theta, \\ \text{s.t.} \quad & \\ \theta \in \mathcal{O}(k), & \end{aligned} \tag{4.104}$$

$$\begin{aligned} \underline{d}_{i+n_\theta}(k) = \arg\min \quad & c_i^{\mathrm{T}}\theta, \\ \text{s.t.} \quad & \\ \theta \in \mathcal{O}(k), & \end{aligned} \tag{4.105}$$

with

$$\mathcal{O}(k) = \bigcap_{i=1}^{2n_\theta} \mathcal{O}_i(k), \tag{4.106}$$

where $\overline{d}_i(k)$ and $\underline{d}_{i+n_\theta}(k)$ are, respectively, the maximal and minimal values of the objective over the bounding set, $\mathcal{O}_i(k)$ denotes the constraint-induced feasible region, and c_i is the i-th column of the objective direction matrix, $i = 1, \dots, n_\theta$. The resulting minimal orthotope is

$$O_k^* = (\bar{\theta}^*(k), d^*(k)), \tag{4.107}$$

where $\bar{\theta}^*(k) = \dfrac{\overline{d}_i(k) + \underline{d}_{i+n_\theta}(k)}{2}$ and $d^*(k) = \dfrac{\overline{d}_i(k) - \underline{d}_{i+n_\theta}(k)}{2}$.

Since the parameters evolve with time, this orthotopic set must be suitably enlarged by modifying the constraints so as to capture the admissible uncertainty range, taking into account the initial parameter set, the measurement strips, the bound on parameter variations, and the expansion direction of the orthotope.

Theorem 4.4 *Consider the identification model in (4.97). Suppose that at time $k-1$ the parameter vector satisfies $\theta(k-1) \in \hat{O}_{k-1} = (\bar{\theta}^*(k-1), d^*(k-1))$ and that the parameter variation obeys $|\theta(k) - \theta(k-1)| \leqslant \gamma$. For the measurement $h(k)$,* *define the strip* $S_k = \{\theta(k) : |h(k) - \varphi^{\mathrm{T}}(k)\theta(k)| \leqslant \sigma\}$*. Then the predicted orthotopic set enclosing the parameter estimates at time k is*

$$\theta(k) \in \hat{O}_k = \left\{\theta(k) : \theta(k) = \overline{\theta}^*(k) + \operatorname{diag}(d^*(k))\omega, \ \|\omega\|_\infty \leqslant 1\right\}, \tag{4.108}$$

where

$$\begin{aligned} \overline{\hat{d}}_i(k) = \arg\max \quad & \hat{c}_i^{\mathrm{T}}\theta, \\ \text{s.t.} \quad & \\ & \theta \in \tilde{O}_k \bigcap S_k, \end{aligned} \tag{4.109}$$

$$\begin{aligned} \underline{\hat{d}}_{i+n_\theta}(k) = \arg\min \quad & \hat{c}_i^{\mathrm{T}}\theta, \\ \text{s.t.} \quad & \\ & \theta \in \tilde{O}_k \bigcap S_k, \end{aligned} \tag{4.110}$$

and

$$\tilde{O}_k = (\tilde{\theta}(k), \tilde{d}(k)), \tag{4.111}$$

$$\tilde{\theta}(k) = \frac{\overline{\tilde{d}}_i(k) + \underline{\tilde{d}}_{i+n_\theta}(k)}{2}, \tag{4.112}$$

$$\tilde{d}(k) = \frac{\overline{\tilde{d}}_i(k) - \underline{\tilde{d}}_{i+n_\theta}(k)}{2}, \tag{4.113}$$

$$\overline{\tilde{d}}_i(k) = d^*(k-1) + \Gamma_k\gamma, \tag{4.114}$$

$$\underline{\tilde{d}}_{i+n_\theta}(k) = d^*(k-1) + (\Gamma_k - I)\gamma, \tag{4.115}$$

$$\Gamma_k = \operatorname{diag}(r_i), \ r_i \in \{0, 1\}, \tag{4.116}$$

where $\gamma = [\gamma_1, \gamma_2, \cdots, \gamma_{n_\theta}]^{\mathrm{T}}$*, and* Γ_k *is a diagonal matrix encoding the expansion direction of the orthotope.*

Proof 4.4. In the LPV model (4.97), each parameter either increases or decreases from time $k-1$ to k. Hence, given the orthotope at time $k-1$, we examine 2^{n_θ} spatial expansion directions to bound the unknown parameters at time k:

$$\tilde{O}_k^{(j)} = (\tilde{\theta}^{(j)}(k), \tilde{d}^{(j)}(k)), \tag{4.117}$$

$$\overline{\tilde{d}}_i^{(j)}(k) = d^*(k-1) + \Gamma_k^{(j)}\gamma, \tag{4.118}$$

$$\underline{\tilde{d}}_{i+n_\theta}^{(j)}(k) = d^*(k-1) + (\Gamma_k^{(j)} - I_{n_\theta})\gamma, \tag{4.119}$$

$$\Gamma_k^{(j)} = \mathrm{diag}(r_i^{(j)}),\ r_i^{(j)} \in \{0, 1\}, \tag{4.120}$$

$$\gamma = [\gamma_1, \gamma_2, \cdots, \gamma_{n\theta}]^{\mathrm{T}}, \tag{4.121}$$

where $i = 1, \dots, n_\theta$, $j = 1, \dots, 2^{n_\theta}$, I_{n_θ} is the identity matrix, and $\Gamma \in \mathbb{R}^{n_\theta \times n_\theta}$ is the coefficient matrix for the expansion direction. Here $\mathrm{diag}(r_i)$ is diagonal with entries in $\{0, 1\}$: if θ_i increases, then $\Gamma_{i,i} = r_i = 1$; otherwise, $\Gamma_{i,i} = r_i = 0$. For $n_\theta = 2$, the four extended spaces are illustrated in Figure 4.4.

Among the 2^{n_θ} directions, only one contains the true parameters. To improve directional accuracy, the correct expansion direction is determined via the intersection between the expanded orthotopes and the measurement strips.

Suppose that, at time $k_\gamma \in \{k, k+1, \cdots, k+n_\gamma\}$, a measurement strip is available and the orthotope is expanded n_γ times. Based on these n_γ expansions and n_γ strips, multiple orthotopic outer bounds for the intersections are obtained:

$$\hat{O}_{k_\gamma}^{(j)} = (\hat{\theta}^{(j)}(k_\gamma), \hat{d}^{(j)}(k_\gamma)), \tag{4.122}$$

$$\begin{aligned} \overline{\hat{d}}_i^{(j)}(k_\gamma) = \arg\max \quad & \hat{c}_{ij}^{\mathrm{T}}\theta, \\ \text{s.t.} \quad & \\ \theta \in \tilde{O}_{k_\gamma}^{(j)} \bigcap S_{k_\gamma} & \end{aligned} \tag{4.123}$$

$$\begin{aligned} \underline{\hat{d}}_i^{(j)}(k_\gamma) = \arg\min \quad & c_{ij}^{\mathrm{T}}\theta. \\ \text{s.t.} \quad & \\ \theta \in \tilde{O}_{k_\gamma}^{(j)} \bigcap S_{k_\gamma}. & \end{aligned} \tag{4.124}$$

The hypervolume associated with the j-th orthotopic outer approximation is defined as:

$$\mathcal{V}^{(j)} * k * \gamma = \prod_{i=1}^{n_\theta} \left(\overline{\hat{d}}^{(j)} * i(k * \gamma) - \underline{\hat{d}}^{(j)} * i(k * \gamma) \right). \tag{4.125}$$

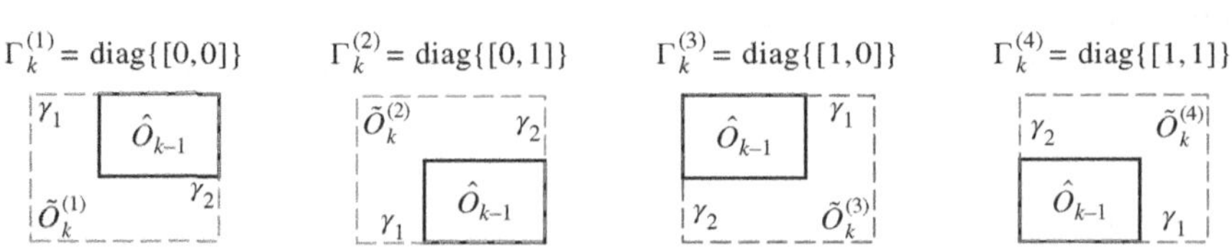

Figure 4.4 Process of orthotope spatial extension when $n_\theta = 2$.

Among all candidate expansions, those that actually enclose the unknown parameters over time usually yield the largest accumulated hypervolume. Therefore, the optimal expansion direction Γ_k is chosen by

$$\Gamma_k = \arg\max \sum_{i=1}^{k_\gamma+1} \mathcal{V}^{(j)} * k * \gamma, \tag{4.126}$$

and the identification orthotope $\hat{O}_k$ at time k is then constructed by intersecting the orthotope $\hat{O}_{k-1}$ (expanded along the correct direction Γ_k) with the measurement strip $S(k)$, yielding (4.108)–(4.116).

Remark 4.1 For discrete-time LPV systems, parameter variations are often unidirectional over a time window. To reduce computation, once Γ_k is determined by (4.126), it can be retained as the expansion direction at times $k, k+1, \cdots, k+n-1$, with the bounding set refined by intersecting with the corresponding strips. If at time $k+n$ an empty intersection occurs, i.e., $\tilde{O}_{k+n} \bigcap S_{k+n} = \emptyset$, this indicates a change in the variation direction. A new Γ_{k+n} is then determined using the strips at $k+n, k+n+1, \cdots, k+n+n_\gamma$.

Remark 4.2 Consider a discrete-time LPV system whose parameters evolve monotonically in a known direction, while the exact variation amplitude is unknown. If two consecutive estimates satisfy $\bar{\theta}^*(k) - \bar{\theta}^*(k-1) < \lambda \ll \gamma$, the identification interval can be further tightened by shrinking the expansion bound in (4.114) and (4.115). In particular, the attenuated bounds are chosen as:

$$\overline{d}_i(k) = d^*(k-1) + \lambda\Gamma_k\gamma, \tag{4.127}$$

$$\tilde{\underline{d}}_{i+n_\theta}(k) = d^*(k-1) + \lambda(\Gamma_k - I_{n_\theta})\gamma, \tag{4.128}$$

where the scalar $\lambda \in (0, 1)$ acts as an attenuation factor that moderates the amount of orthotope expansion.

Remark 4.3 Based on (4.122)–(4.126), let us illustrate the mechanism for selecting the expansion direction by considering the simple case $n_\theta = 2$ and $n_\gamma = 2$. Figure 4.5 depicts the procedure: the solid rectangle represents the orthotopic set $\hat{O}_{k-1}$ at time instant $k-1$. At times $k+n, k+n+1$, and $k+n+2$, the orthotope is propagated in four candidate directions and, in each case, intersected with the corresponding measurement strips. Panel (a) shows the geometric expansion of $\hat{O}_{k-1}$ when the direction matrix is chosen as $\Gamma = \mathrm{diag}([0, 0])$; under this choice, the predicted orthotopes used for time-varying parameter identification—subject to the strip constraints—are indicated by the dashed rectangles. The hypervolume of the outer orthotope in each candidate direction is evaluated via (4.125). Similarly, panels (b)–(d) present the expansions along the remaining directions and the associated hyperspaces. According to the selection rule in (4.126), one obtains

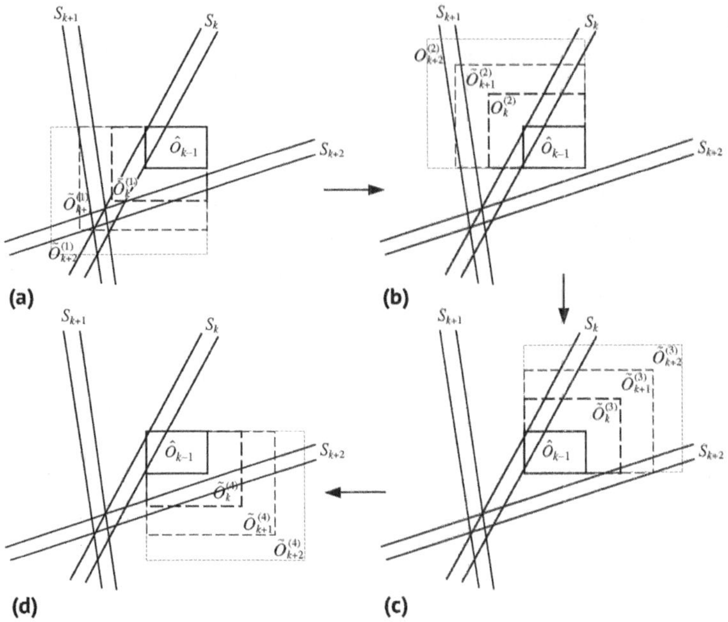

Figure 4.5 Process of determining the orthotope expansion direction.

that $\tilde{O}_k^{(1)}$ is the orthotope that contains the true parameters, and the corresponding expansion direction at time k is $\Gamma = \text{diag}([0, 0])$.

4.5.2.2 Optimal OSE-EZKF Design for the Discrete-time LPV System

Based on the orthotopic identification of the time-varying parameters in (4.97), the uncertain system matrices $A(\theta(k))$ and $B(\theta(k))$ can be embedded in orthotopic sets as follows:

$$\begin{aligned} A(\theta(k)) \in \mathcal{A}_k &= \{A(k) \in \mathbb{R}^{n_x \times n_x} : A(k) = \bar{A}(k) + \text{diag}(d_A(k))\omega,\ \|\omega\|_\infty \leqslant 1\} \\ &= \{A(k) \in \mathbb{R}^{n_x \times n_x} : A(k) = \bar{A}(k) + \Delta A(k),\ \Delta A(k) = \text{diag}(d_A(k))\omega\} \\ &= (\bar{A}(k), d_A(k)), \end{aligned} \tag{4.129}$$

$$\begin{aligned} B(\theta(k)) \in \mathcal{B}_k &= \{B(k) \in \mathbb{R}^{n_u} : B(k) = \bar{B}(k) + \text{diag}(d_B(k))\omega,\ \|\omega\|_\infty \leqslant 1\} \\ &= \{B(k) \in \mathbb{R}^{n_u} : B(k) = \bar{B}(k) + \Delta B(k),\ \Delta B(k) = \text{diag}(d_B(k))\omega\} \\ &= (\bar{B}(k), d_B(k)). \end{aligned} \tag{4.130}$$

With these orthotopic bounds on the parameter matrices of the LPV model (4.96) at hand, we proceed to design the state estimator. The state equation can be rewritten as:

$$\begin{cases} x(k+1) = (\bar{A}(k) + \Delta A(k))x(k) + (\bar{B}(k) + \Delta B(k))u(k) + Ew(k), \\ y(k) = Cx(k) + Fv(k), \end{cases} \tag{4.131}$$

where $\bar{A}(k)$ is the identified nominal matrix and $\Delta A(k)$ captures the parametric uncertainty.

For the uncertain discrete system (4.131), the proposed OSE-EZKF is constructed on the following observer:

$$\begin{aligned} \hat{x}(k+1) &= (\bar{A}(k) + \Delta A(k))\hat{x}(k) + (\bar{B}(k) + \Delta B(k))u(k) + Ew(k) \\ &\quad + L(k)\big(y(k) - C\hat{x}(k) - Fv(k)\big). \end{aligned} \tag{4.132}$$

Theorem 4.5 *Suppose that, at time k, the state satisfies* $x(k) \in \hat{X}_k = \langle \hat{p}(k), \hat{G}(k) \rangle$, *with* $\hat{p}(k) \in \mathbb{R}^{n_x}$ *and* $\hat{G}(k) \in \mathbb{R}^{n_x \times s}$. *Then, at time* $k+1$, *the estimated state is enclosed by the zonotope*

$$\hat{x}(k+1) \in \hat{X}(k+1) = \langle \hat{p}(k+1), \hat{G}(k+1) \rangle, \tag{4.133}$$

$$\hat{p}(k+1) = (\bar{A}(k) - L(k)C)\hat{p}(k) + \bar{B}(k)u(k) + L(k)y(k), \tag{4.134}$$

$$\hat{G}(k+1) = [(\bar{A}(k) - L(k)C)\hat{G}(k)\; M_1(k)\; M_2(k)\; EG_w \;\; -L(k)FG_v], \tag{4.135}$$

where $M_1 \in \mathbb{R}^{n_x}$ *and* $M_2 \in \mathbb{R}^{n_u}$ *are column vectors determined by* $\Delta A(k)$ *and* $\Delta B(k)$, *respectively, and satisfy*

$$M_1(i) = |\phi_1(i,j)|, \qquad i = 1, \dots, n_x, \tag{4.136}$$

$$M_2(i) = |\phi_2(i)|, \qquad i = 1, \dots, n_u, \tag{4.137}$$

with $\phi_1 = [d_A(k)p(k)]^+ + [d_A(k)G_{\downarrow_1}(k)]^+$ *and* $\phi_2 = [d_B(k)u(k)]^+$.

Proof 4.5. From (4.131) and zonotope properties, the OSE-EZKF update reads

$$\begin{aligned} \hat{x}(k+1) &= (\bar{A}(k) + \Delta A(k))\hat{x}(k) + (\bar{B}(k) + \Delta B(k))u(k) + Ew(k) \\ &\quad + L(k)\big(y(k) - C\hat{x}(k) - Fv(k)\big) \\ &= (\bar{A}(k) + \Delta A(k) - L(k)C)\hat{x}(k) + (\bar{B}(k) + \Delta B(k))u(k) \\ &\quad + Ew(k) + L(k)y(k) - L(k)Fv(k) \\ &\in \big((\bar{A}(k) + \Delta A(k) - L(k)C) \odot \langle \hat{p}(k), \hat{G}(k) \rangle\big) \oplus \big((\bar{B}(k) + \Delta B(k))u(k)\big) \\ &\quad \oplus \big(E \odot \langle 0, G_w(k) \rangle\big) \oplus \big(L(k)y(k)\big) \oplus \big(-L(k)F \odot \langle 0, G_v(k) \rangle\big), \end{aligned} \tag{4.138}$$

where

$$
\begin{aligned}
&(\bar{A}(k) + \Delta A(k)) \odot \langle \hat{p}(k), \hat{G}(k) \rangle \\
&\quad = \bar{A}(k)\hat{p}(k) \oplus \bar{A}(k)\hat{G}(k)\mathbf{B}^s \oplus \left(\Delta A(k) \odot \langle \hat{p}(k), \hat{G}(k) \rangle \right),
\end{aligned} \tag{4.139}
$$

$$
(\bar{B}(k) + \Delta B(k))u(k) = \bar{B}(k)u(k) + \Delta B(k)u(k). \tag{4.140}
$$

From (4.129) to (4.130),

$$
\Delta A(k) = \mathrm{diag}(d_A(k))\omega, \quad \Delta B(k) = \mathrm{diag}(d_B(k))\omega, \quad \|\omega\|_\infty \leqslant 1. \tag{4.141}
$$

By Definition 4.3, the generator matrix satisfies

$$
\hat{p}(k) \oplus \hat{G}(k)\mathbf{B}^s \subseteq \hat{p}(k) \oplus \hat{G}(k)_{\downarrow_1}(k)\mathbf{B}^1. \tag{4.142}
$$

Hence,

$$
\Delta A(k) \odot \langle \hat{p}(k), \hat{G}(k) \rangle \subseteq M_1(k)\mathbf{B}^1, \tag{4.143}
$$

$$
\begin{aligned}
M_1(k) &= [\max(\Delta A(k))\hat{p}(k)]^+ + [\max(\Delta A(k))\hat{G}_{\downarrow_1}(k)]^+ \\
&= [d_A(k)\hat{p}(k)]^+ + [d_A(k)\hat{G}_{\downarrow_1}(k)]^+,
\end{aligned} \tag{4.144}
$$

$$
\Delta B(k)u(k) \subseteq M_2(k)\mathbf{B}^1, \tag{4.145}
$$

$$
M_2(k) = [\max(\Delta B(k))u(k)]^+ = [d_B(k)u(k)]^+, \tag{4.146}
$$

where $M_1(k) \in \mathbb{R}^{n_x}$ and $M_2(k) \in \mathbb{R}^{n_u}$ are column vectors.
Substituting (4.139)–(4.146) into (4.138) yields

$$
\begin{aligned}
\hat{x}(k+1) \in \hat{X}(k+1) &= \langle \hat{p}(k+1), \hat{G}(k+1) \rangle \\
&= [(\bar{A}(k) - L(k)C)\hat{p}(k) + \bar{B}(k)u(k) + L(k)y(k)] \\
&\quad \oplus [(\bar{A}(k) - L(k)C)\hat{G}(k) \;\; M_1(k) \;\; M_2(k) \;\; EG_w \;\; -L(k)FG_v]\mathbf{B}^{s+n_x+n_y+2}.
\end{aligned}
$$

According to the above derivation, each iteration increases the order of the zonotopic set $\hat{X}$ by $n_x + n_y + 2$. As the iterations proceed, this order may become computationally intractable. To control complexity, we apply the dimensionality reduction operator to $\hat{G}(k+1)$; then, by (4.133),

$$
\hat{X}(k+1) = \langle \hat{p}(k+1), \hat{G}(k+1) \rangle \subseteq \langle \hat{p}(k+1), \hat{G}_{\downarrow_s}(k+1) \rangle. \tag{4.147}
$$

To obtain a tight zonotopic enclosure, the gain $L(k)$ is selected so as to minimize the Frobenius radius of $\hat{X}(k+1)$. Invoking Definitions 4.3–4.7, this leads to the following result.

Theorem 4.6 *For the zonotope in* (4.133), *minimizing* $\|\langle\hat{p}(k+1), \hat{G}(k+1)\rangle\|_{F,M}$ *yields*

$$L(k) = \bar{A}(k)\hat{G}(k)\hat{G}^{\mathrm{T}}(k)C^{\mathrm{T}}\left[C\hat{G}(k)\hat{G}^{\mathrm{T}}(k)C^{\mathrm{T}} + FG_v G_v^{\mathrm{T}}F^{\mathrm{T}}\right]^{-1}. \tag{4.148}$$

Proof 4.6. By Definition 4.5, the F-radius of $\hat{X}(k+1)$ is

$$\|\hat{X}(k+1)\|_{F,M} = \|\hat{G}(k+1)\|_{F,M} = \sqrt{\mathrm{tr}\left(\hat{G}^{\mathrm{T}}(k+1)M\hat{G}(k+1)\right)}. \tag{4.149}$$

Let $J_{F,M}(k+1) = \|\hat{G}(k+1)\|^2_{F,M} = \mathrm{tr}\left(\hat{G}^{\mathrm{T}}(k+1)M\hat{G}(k+1)\right)$. Then

$$\begin{aligned}
J_{F,M} &= \mathrm{tr}\Big[M(\bar{A}(k) - L(k)C)\hat{G}(k)\hat{G}^{\mathrm{T}}(k)(\bar{A}(k) - L(k)C)^{\mathrm{T}} + MM_1M_1^{\mathrm{T}} \\
&\quad + MM_2M_2^{\mathrm{T}} + MEG_wG_w^{\mathrm{T}}E^{\mathrm{T}} + ML(K)FH_vH_v^{\mathrm{T}}F^{\mathrm{T}}L^{\mathrm{T}}(k)\Big] \\
&= \mathrm{tr}\Big[M\big(L(k)C\hat{G}(k)\hat{G}^{\mathrm{T}}(k)C^{\mathrm{T}}L^{\mathrm{T}}(k) - L(k)C\hat{G}(k)\hat{G}^{\mathrm{T}}(k)\bar{A}^{\mathrm{T}}(k) \\
&\quad - \bar{A}(k)\hat{G}(k)\hat{G}^{\mathrm{T}}(k)C^{\mathrm{T}}L^{\mathrm{T}}(k) + L(K)FH_vH_v^{\mathrm{T}}F^{\mathrm{T}}L^{\mathrm{T}}(k)\big)\Big] \\
&\quad + \mathrm{tr}\Big[M\big(M_1M_1^{\mathrm{T}} + M_2M_2^{\mathrm{T}} + MEG_wG_w^{\mathrm{T}}E^{\mathrm{T}}\big)\Big].
\end{aligned} \tag{4.150}$$

Setting $\partial_{L(k)}J_{F,M}(k+1) = 0$ gives the minimizing condition (hence the minimum F-radius and an optimal zonotope), namely

$$\begin{aligned}
&\partial_{L(k)}\, J_{F,M}(k+1) \\
&\quad = \partial_{L(k)}\mathrm{tr}\Big[M\big(L(k)C\hat{G}(k)\hat{G}^{\mathrm{T}}(k)C^{\mathrm{T}}L^{\mathrm{T}}(k) + L(K)FH_vH_v^{\mathrm{T}}F^{\mathrm{T}}L^{\mathrm{T}}(k) \\
&\quad -2\bar{A}(k)\hat{G}(k)\hat{G}^{\mathrm{T}}(k)C^{\mathrm{T}}L^{\mathrm{T}}(k)\big)\Big] = 0,
\end{aligned} \tag{4.151}$$

$$\begin{aligned}
&\partial_{L(k)}\mathrm{tr}\Big[M\big(L(k)C\hat{G}(k)\hat{G}^{\mathrm{T}}(k)C^{\mathrm{T}}L^{\mathrm{T}}(k) + L(K)FH_vH_v^{\mathrm{T}}F^{\mathrm{T}}L^{\mathrm{T}}(k)\big)\Big] \\
&= 2\,\partial_{L(k)}\mathrm{tr}\Big[M\big(\bar{A}(k)\hat{G}(k)\hat{G}^{\mathrm{T}}(k)C^{\mathrm{T}}L^{\mathrm{T}}(k)\big)\Big],
\end{aligned} \tag{4.152}$$

where M is a symmetric positive definite weighting matrix. Hence,

$$L(k)[C\hat{G}(k)\hat{G}^{\mathrm{T}}(k)C^{\mathrm{T}} + FH_vH_v^{\mathrm{T}}F^{\mathrm{T}}] = \bar{A}(k)\hat{G}(k)\hat{G}^{\mathrm{T}}(k)C^{\mathrm{T}}, \tag{4.153}$$

and, since $L(k)$ is assumed to be full rank, it follows that

$$L(k) = \bar{A}(k)\hat{G}(k)\hat{G}^{\mathrm{T}}(k)C^{\mathrm{T}}[C\hat{G}(k)\hat{G}^{\mathrm{T}}(k)C^{\mathrm{T}} + FG_vG_v^{\mathrm{T}}F^{\mathrm{T}}]^{-1}. \tag{4.154}$$

Given the optimal zonotope $\hat{X}(k+1) = \langle \hat{p}(k+1), \hat{G}_{\downarrow s}(k+1) \rangle$, the true state $x(k+1)$ is contained in the axis-aligned box $\square \hat{X}(k+1) = [\underline{\hat{x}}(k+1), \overline{\hat{x}}(k+1)]$, whose components are

$$\underline{\hat{x}}(k+1) = \hat{p}_i(k+1) - \sum_{j=1}^{s} \left|\hat{G}_{\downarrow s}(k+1)\right|_{i,j}, \quad i = 1, \dots, n_x, \tag{4.155}$$

$$\overline{\hat{x}}(k+1) = \hat{p}_i(k+1) + \sum_{j=1}^{s} \left|\hat{G}_{\downarrow s}(k+1)\right|_{i,j}, \quad i = 1, \dots, n_x, \tag{4.156}$$

where $\underline{\hat{x}}(k+1)$ and $\overline{\hat{x}}(k+1)$ denote, respectively, the lower and upper bounds of $x(k+1)$.

4.5.3 Conclusion

This section presented a state-estimation framework for discrete-time LPV systems affected simultaneously by stochastic disturbances and UBB uncertainties. The method combines an orthotope–search–expansion (OSE) mechanism with an EZKF. The OSE step exploits a sequence of measurement strips to iteratively adjust an orthotopic region in the parameter space, yielding compact bounds on the time-varying parameter vector, which are then mapped to the uncertain matrices $A(\theta(k))$ and $B(\theta(k))$. Within this bounded-parameter description, the EZKF augments the classical Kalman recursion with a zonotopic enclosure, so that both covariance information and set-membership bounds are propagated in time. By minimizing the Frobenius radius of the predicted zonotope, an optimal gain matrix is obtained that balances the impact of stochastic noise and UBB disturbances, leading to tight state intervals and accurate estimates. The resulting algorithm is suitable for real-time implementation and shows clear robustness advantages over purely stochastic or purely set-membership filters. Beyond the LPV case studied here, the same OSE-EZKF idea can be extended to applications such as electric power systems, multi-sensor security-fusion estimation, and ground vehicle dynamics, and, with zonotopic linearization, to nonlinear systems operating under mixed uncertainties.

4.6 Constrained Zonotopic Kalman Filtering

4.6.1 Preliminaries

In this subsection, we briefly recall the notation related to zonotopes and introduce some basic concepts of difference-of-convex (DC) programming, which will be used to formulate the gain computation of the constrained zonotopic Kalman filter as a DC optimization problem. And the fundamental definitions and properties of zonotopes, including their representation, Minkowski sums, and linear images, have already been introduced in the previous subsections. In what follows, we will directly use this notation and only recall additional results when necessary.

Based on the aforementioned zonotopic notation, the basic notions of DC programming are introduced as follows.

Definition 4.8 A real-valued mapping $f : X \to \mathbb{R}$, with $X \subseteq \mathbb{R}^n$, is said to be a DC function on X if there exist two convex functions $g, h : X \to \mathbb{R}$ such that $f(x) = g(x) - h(x)$ for all $x \in X$.

Definition 4.9 The subdifferential of a convex function $f : X \to \mathbb{R}$ at a point $x_0 \in X$ is defined by

$$\partial f(x_0) = \left\{ u_0 \in \mathbb{R}^n : f(x) \geq f(x_0) + u_0^{\mathrm{T}}(x - x_0),\ \forall x \in X \right\},$$

where, in the differentiable case, $u_0 = (\partial/\partial x) f(x_0)$ coincides with the usual gradient of f at x_0.

Assume that the initial state and the UBB noises in system (4.7) belong to the zonotopic sets

$$x_0 \in \mathcal{X}_0 = p_0 \oplus G_0 \mathbf{B}^s = \langle p_0, G_0 \rangle, \qquad w_k \in \mathcal{W} = \langle 0, W \rangle, \qquad v_{k+1} \in \mathcal{V} = \langle 0, V \rangle,$$

with $G_0 = \mathrm{diag}(\tilde{x}_0)$, $W = \mathrm{diag}(\tilde{w})$, and $V = \mathrm{diag}(\tilde{v})$. For the nonlinear system in (4.7), the state trajectory is further constrained by

$$\|l(x_k)\| \leqslant \xi, \qquad l(x_k) = D_k - L_k x_k. \tag{4.157}$$

Since the dynamics must evolve inside the constraint set (4.157), a central problem is how to exploit this information to contract the feasible state region encoded by the above zonotopes. The aim of this work is therefore to construct a zonotopic filtering scheme that performs state estimation for nonlinear systems while rigorously incorporating such state constraints.

4.6.2 Main Results

4.6.2.1 Bounding the Linearization Error

Classical linearization via Taylor expansion requires continuous differentiability and explicit Jacobian/Hessian evaluations at each point, which can be computationally demanding. In this work, the Stirling interpolation formula is employed to approximate the nonlinear mappings. For the nonlinearity $f(x_k)$ in (4.7), we write

$$\begin{aligned} f(x_k) &= f(\hat{x}_k) + F_k(x_k - \hat{x}_k) + H.O.T_1 \\ &= f_L(x_k) + H.O.T_1, \end{aligned} \tag{4.158}$$

$$\begin{aligned} h(x_{k+1}) &= h(\hat{x}_{k+1,k}) + H_{k+1}(x_{k+1} - \hat{x}_{k+1,k}) + H.O.T_2 \\ &= h_L(x_{k+1}) + H.O.T_2, \end{aligned} \tag{4.159}$$

where

$$F_k = \frac{1}{2l}\begin{bmatrix} \left(f(\hat{x}_k + le_1) - f(\hat{x}_k - le_1)^{\mathrm{T}}\right) \\ \left(f(\hat{x}_k + le_2) - f(\hat{x}_k - le_2)^{\mathrm{T}}\right) \\ \vdots \\ \left(f(\hat{x}_k + le_n) - f(\hat{x}_k - le_n)^{\mathrm{T}}\right) \end{bmatrix}^{\mathrm{T}}, \tag{4.160}$$

$$H_k = \frac{1}{2l}\begin{bmatrix} \left(h(\hat{x}_{k+1,k} + le_1) - h(\hat{x}_{k+1,k} - le_1)^{\mathrm{T}}\right) \\ \left(h(\hat{x}_{k+1,k} + le_2) - h(\hat{x}_{k+1,k} - le_2)^{\mathrm{T}}\right) \\ \vdots \\ \left(h(\hat{x}_{k+1,k} + le_n) - h(\hat{x}_{k+1,k} - le_n)^{\mathrm{T}}\right) \end{bmatrix}^{\mathrm{T}}, \tag{4.161}$$

$H.O.T_1$ and $H.O.T_2$ denote higher-order terms arising from linearization, l is the finite-difference step, $\hat{x}_k$ denotes the a posteriori state estimate at time k, while $\hat{x}_{k+1,k}$ represents the corresponding a priori prediction. The vectors $e_i \in \mathbb{R}^n$, $i = 1, \dots, n$, are the canonical basis of $\mathbb{R}^n$, i.e., $e_1 = [1, 0, \dots, 0]^{\mathrm{T}}_{n\times 1}$, $e_2 = [0, 1, \dots, 0]^{\mathrm{T}}_{n\times 1}$, ..., $e_n = [0, 0, \dots, 1]^{\mathrm{T}}_{n\times 1}$.

To bound the linearization errors, we employ a nonlinear programming (NLP) approach to obtain compact interval enclosures. From (4.158) to (4.159), the errors read

$$\Delta f(x_k) = f(x_k) - f_L(x_k), \tag{4.162}$$

$$\Delta h(x_{k+1}) = h(x_{k+1}) - h_L(x_{k+1}). \tag{4.163}$$

Their ranges are represented as interval vectors

$$H.O.T_1 = \Delta f(x_k) = [\Delta f_{\min}(x_k), \Delta f_{\max}(x_k)], \tag{4.164}$$

$$H.O.T_2 = \Delta h(x_{k+1}) = [\Delta h_{\min}(x_{k+1}), \Delta h_{\max}(x_{k+1})]. \tag{4.165}$$

We construct the following NLP to compute the linearization-error bounds:

$$\begin{aligned} &\min\ \Delta f_k^i \\ &\text{s.t. } x_k^i \in X_k^i, \quad i = 1, \dots, n, \end{aligned} \tag{4.166}$$

where the superscript i indexes the i-th component of the state vector, and X_k denotes the feasible set of x_k, which is enclosed by the zonotope $\mathcal{X}_{k+1}$.

To recast the above problem within a DC programming framework, the nonlinearity $f(x_k)$ is written as a DC functions, $f(x_k) = g_1(x_k) - g_2(x_k)$, where $g_1(x_k) = f(x_k) + \alpha x_k^{\mathrm{T}} x_k$ and $g_2(x_k) = \alpha x_k^{\mathrm{T}} x_k$ with $\alpha > 0$. The subgradients of g_1 and g_2 at $\hat{x}_k$ are

$$u_1 = \frac{\partial g_1(\hat{x}_k)}{\partial x}, \quad u_2 = \frac{\partial g_2(\hat{x}_k)}{\partial x}. \tag{4.167}$$

By convex subdifferential calculus, the linear under-estimators satisfy

$$g_1(x_k) \geq \bar{g}_1(x_k), \qquad g_2(x_k) \geq \bar{g}_2(x_k), \tag{4.168}$$

with

$$\bar{g}_1(x_k) = g_1(\hat{x}_k) + u_1^{\mathrm{T}}(x_k - \hat{x}_k), \tag{4.169}$$

$$\bar{g}_2(x_k) = g_2(\hat{x}_k) + u_2^{\mathrm{T}}(x_k - \hat{x}_k). \tag{4.170}$$

Hence, the linearization-error bounds can be written as:

$$\begin{aligned} \Delta f_{\min}^{(i)}(x_k) &= \min_{x_k \in X_k} \left\{ g_1^{(i)}(x_k) - \bar{g}_2^{(i)}(x_k) - f_L^{(i)}(x_k) \right\}, \\ \Delta f_{\max}^{(i)}(x_k) &= \max_{x_k \in X_k} \left\{ \bar{g}_1^{(i)}(x_k) - g_2^{(i)}(x_k) - f_L^{(i)}(x_k) \right\}. \end{aligned} \tag{4.171}$$

Consequently, global lower and upper bounds are obtained by evaluating the function at every vertex of the zonotope $\mathcal{Z}_{F,k}$. We then compute the center and shape matrix of the zonotope $\mathcal{Z}_{F,k}$ that bounds the linearization error:

$$\begin{aligned} p_{F,k} &= \left\{ \frac{\Delta f_{\min}^{(1)} + \Delta f_{\max}^{(1)}}{2}, \frac{\Delta f_{\min}^{(2)} + \Delta f_{\max}^{(2)}}{2}, \dots, \right. \\ &\qquad \left. \frac{\Delta f_{\min}^{(n)} + \Delta f_{\max}^{(n)}}{2} \right\}^{\mathrm{T}}, \\ G_{F,k} &= \mathrm{diag} \left\{ \frac{\Delta f_{\max}^{(1)} - \Delta f_{\min}^{(1)}}{2}, \frac{\Delta f_{\max}^{(2)} - \Delta f_{\min}^{(2)}}{2}, \dots, \right. \\ &\qquad \left. \frac{\Delta f_{\max}^{(n)} - \Delta f_{\min}^{(n)}}{2} \right\}. \end{aligned} \tag{4.172}$$

4.6.2.2 Prediction Step

After the linearization procedure, the state dynamics of the nonlinear system (4.7) can be rewritten as:

$$x_{k+1} = f_L(x_k) + H.O.T_1 + w_k, \tag{4.173}$$

where $f_L(x_k)$ denotes the linearized component of $f(x_k)$ and can be regarded as a linear affine transformation of the feasible state set X_k, i.e., $f_L(x_k) \in \mathcal{Z}_{L,k}$. In addition, the higher-order term satisfies $H.O.T_1 \in \mathcal{Z}_{F,k} = \langle p_{F,k}, G_{F,k} \rangle$ and the process disturbance lies in $w_k \in \mathcal{W} = \langle 0, W \rangle$. Consequently, the prediction $\hat{x}_{k+1,k}$ belongs to

$$\hat{x}_{k+1,k} \in \bar{\mathcal{X}}_{k+1} = \mathcal{Z}_{L,k} \oplus \mathcal{Z}_{F,k} \oplus \mathcal{W}. \tag{4.174}$$

Theorem 4.7 *Given $x_k \in \mathcal{X}_k = \langle \hat{p}_k, \hat{G}_k \rangle$, the one-step prediction at time $k+1$ is enclosed by the zonotope*

$$\hat{x}_{k+1,k} \in \bar{\mathcal{X}}_{k+1} = \langle \bar{p}_{k+1}, \bar{G}_{k+1} \rangle,$$

where

$$\bar{p}_{k+1} = f(\hat{p}_k) + p_{F,k}, \tag{4.175}$$

$$\overline{G}_{k+1} = \begin{bmatrix} F_k \hat{G}_k & G_{F,k} & W \end{bmatrix}. \tag{4.176}$$

Proof 4.7. Using standard properties of zonotopes, one obtains

$$\begin{aligned}
\hat{x}_{k+1,k} &\in \mathcal{Z}\left(\overline{p}_{k+1}, \overline{G}_{k+1}\right) = \mathcal{Z}_{L,k} \oplus \mathcal{Z}_{F,k} \oplus \mathcal{W} \\
&= \left(f(\hat{p}_k) + [F_k \otimes \langle \hat{p}_k, \hat{G}_k \rangle] - F_k \hat{p}_k\right) \oplus \langle p_{F,k}, G_{F,k} \rangle \oplus \langle 0, W \rangle \\
&= f(\hat{p}_k) \oplus \langle 0, F_k \hat{G}_k \rangle \oplus \langle p_{F,k}, G_{F,k} \rangle \oplus \langle 0, W \rangle \\
&= \left\langle f(\hat{p}_k) + p_{F,k}, \begin{bmatrix} F_k \hat{G}_k \; G_{F,k} \; W \end{bmatrix} \right\rangle = \langle \overline{p}_{k+1}, \overline{G}_{k+1} \rangle.
\end{aligned}$$

4.6.2.3 Update Step

After linearization, the measurement equation associated with (4.7) can be written as:

$$y_{k+1} = h_L(x_{k+1}) + H.O.T_2 + v_{k+1}, \tag{4.177}$$

where $h_L(x_{k+1})$ denotes the linearized component of $h(x_{k+1})$, the higher-order term satisfies $H.O.T_2 \in [\Delta h_{\min}(x_{k+1}), \Delta h_{\max}(x_{k+1})]$, and $v_{k+1} \in \mathcal{V} = \langle 0, V \rangle$ is the bounded measurement noise.

To embed the state constraint into the measurement model, we augment (4.177) as:

$$\begin{aligned}
\overline{y}_{k+1} = &\begin{bmatrix} h(\hat{x}_{k+1,k}) \\ 0_d \end{bmatrix} + \overline{H}_{k+1} x_{k+1} - \begin{bmatrix} H_{k+1} \hat{x}_{k+1,k} \\ 0_d \end{bmatrix} \\
&+ \begin{bmatrix} H.O.T_2 \\ 0_d \end{bmatrix} + \overline{v}_{k+1},
\end{aligned} \tag{4.178}$$

with

$$\bar{y}_{k+1} = \begin{bmatrix} y_{k+1} \\ D_{k+1} \end{bmatrix}, \qquad \bar{H}_{k+1} = \begin{bmatrix} H_{k+1} \\ L_{k+1} \end{bmatrix}, \qquad \bar{v}_{k+1} = \begin{bmatrix} v_{k+1} \\ \xi_{k+1} \end{bmatrix},$$

where $|\bar{v}_k| \leqslant \tilde{\bar{v}}$, $\bar{V} = \text{diag}(\tilde{\bar{v}})$, $\xi_{k+1} \in \langle 0, \xi \cdot I_{p\times d} \rangle$, and $D_k \in \mathbb{R}^d$.

As a consequence, the state x_{k+1} must satisfy the strip constraint $\mathcal{S}$ given by

$$S = \left\{ \begin{array}{l} x_{k+1} \in \mathbb{R}^n : \\ \left| \overline{H}_{k+1} x_{k+1} - \overline{y}_{k+1} - \begin{bmatrix} H_{k+1}\hat{x}_{k+1,k} \\ 0_d \end{bmatrix} + \begin{bmatrix} h(\hat{x}_{k+1,k}) \\ 0_d \end{bmatrix} + \begin{bmatrix} H.O.T_2 \\ 0_d \end{bmatrix} \right| \leqslant \tilde{\overline{v}} \end{array} \right\}.$$

Using the bounds $H.O.T_2 \in [\Delta h_{\min}(x_{k+1}), \Delta h_{\max}(x_{k+1})]$, the strip $\mathcal{S}$ can be equivalently rewritten as:

$$\mathcal{S} = \left\{ \begin{array}{l} x_{k+1} \in \mathbb{R}^n : \\ \left| \overline{H}_{k+1} x_{k+1} - \overline{y}_{k+1} - \begin{bmatrix} H_{k+1}\hat{x}_{k+1,k} \\ 0_d \end{bmatrix} + \begin{bmatrix} h(\hat{x}_{k+1,k}) \\ 0_d \end{bmatrix} + \begin{bmatrix} s \\ 0_d \end{bmatrix} \right| \leqslant \begin{bmatrix} \tilde{v} + \sigma \\ \xi \end{bmatrix} \end{array} \right\},$$

where

$$s = \frac{\Delta h_{\max}(x_{k+1}) + \Delta h_{\min}(x_{k+1})}{2}, \qquad \sigma = \frac{\Delta h_{\max}(x_{k+1}) - \Delta h_{\min}(x_{k+1})}{2}.$$

By intersecting the prediction zonotope with this strip, we obtain $x_{k+1} \in \bar{\mathcal{X}}_{k+1} \cap \mathcal{S}$. According to Lemma, this intersection is outer-bounded by an updated zonotope $\mathcal{X}_{k+1}(\lambda_{k+1}) = \langle \hat{p}_{k+1}(\lambda_{k+1}) \hat{G}_{k+1}(\lambda_{k+1}) \rangle$, referred to as the update zonotope of x_{k+1}, where

$$\begin{aligned} \hat{p}_{k+1} = {} & \left(I + \lambda_{k+1}\overline{H}_{k+1}\right)^{-1} \left(\overline{p}_{k+1} + \lambda_{k+1}\left(\overline{y}_{k+1} \right.\right. \\ & \left.\left. - \begin{bmatrix} h(\hat{x}_{k+1,k}) \\ 0_d \end{bmatrix} + \begin{bmatrix} H_{k+1}\hat{x}_{k+1,k} \\ 0_d \end{bmatrix} - \begin{bmatrix} s \\ 0_d \end{bmatrix} \right)\right), \end{aligned} \tag{4.179}$$

$$\hat{G}_{k+1} = [(I_{n+d} - \lambda_{k+1}\bar{H}_{k+1})\bar{G}_{k+1} \quad \lambda_{k+1} D_v], \tag{4.180}$$

with $D_v = \text{diag}\left(\begin{bmatrix} \tilde{v} + \sigma \\ \xi \end{bmatrix}\right)$ and λ_{k+1} denoting the filter-gain matrix.

To design λ_{k+1}^*, we adopt the Frobenius norm of the generator $\hat{G}_{k+1}(\lambda_{k+1})$ as an optimality criterion, so that the "size" of the updated zonotope $\mathcal{X}_{k+1}(\lambda_{k+1})$ is minimized. The resulting optimal gain is characterized in the following.

Theorem 4.8 *For x_{k+1} with*

$$x_{k+1} \in \mathcal{X}_{k+1}(\lambda_{k+1}) = \langle \hat{p}_{k+1}(\lambda_{k+1}), \hat{G}_{k+1}(\lambda_{k+1}) \rangle,$$

the optimal filter gain is

$$\lambda_{k+1}^* = \bar{G}_{k+1}\bar{G}_{k+1}^{\mathrm{T}}\bar{\mathrm{H}}_{k+1}^{\mathrm{T}}\left(\bar{H}_{k+1}\bar{G}_{k+1}\bar{G}_{k+1}^{\mathrm{T}}\bar{\mathrm{H}}_{k+1}^{\mathrm{T}} + D_v D_v^{\mathrm{T}}\right)^{-1}. \tag{4.181}$$

Proof 4.8. Decompose $\hat{G}_{k+1}(\lambda_{k+1})$ as

$$\hat{G}_{k+1}(\lambda_{k+1}) = M + \lambda_{k+1} a^{\mathrm{T}}, \tag{4.182}$$

where

$$M = \begin{bmatrix} \bar{G}_{k+1} & 0_{(n+d)\times(v+d)} \end{bmatrix}, \qquad a^{\mathrm{T}} = \begin{bmatrix} -\bar{H}_{k+1}\bar{G}_{k+1} & D_v \end{bmatrix}.$$

Then

$$\begin{aligned} \left\|\hat{G}_{k+1}(\lambda_{k+1})\right\|_F^2 &= \|M + \lambda_{k+1} a^{\mathrm{T}}\|_F^2 \\ &= \mathrm{tr}\left((M^{\mathrm{T}} + a\lambda_{k+1}^{\mathrm{T}})(M + \lambda_{k+1} a^{\mathrm{T}})\right) \\ &= \mathrm{tr}\left(M^{\mathrm{T}} M\right) + 2\,\mathrm{tr}\left(M^{\mathrm{T}}\lambda_{k+1} a^{\mathrm{T}}\right) + \mathrm{tr}\left(a\lambda_{k+1}^{\mathrm{T}}\lambda_{k+1} a^{\mathrm{T}}\right), \end{aligned} \tag{4.183}$$

where $\mathrm{tr}(\cdot)$ denotes the matrix trace and $\mathrm{tr}(XY) = \mathrm{tr}(Y^{\mathrm{T}}X^{\mathrm{T}})$. Minimization with respect to λ_{k+1} yields

$$\lambda_{k+1}^* = -Ma(a^{\mathrm{T}}a)^{-1} = \bar{G}_{k+1}\bar{G}_{k+1}^{\mathrm{T}}\bar{H}_{k+1}^{\mathrm{T}}\left(\bar{H}_{k+1}\bar{G}_{k+1}\bar{G}_{k+1}^{\mathrm{T}}\bar{H}_1^{\mathrm{T}} + D_v D_v^{\mathrm{T}}\right)^{-1}.$$

4.6.3 Conclusion

This section has developed a zonotopic Kalman filtering framework for nonlinear systems subject to explicit state constraints. The method first employs Stirling interpolation to obtain a locally linear representation of the nonlinear dynamics and lifts the state constraints into the output space of the linearized model. The uncertainty caused by linearization is quantified via DC programming, and the resulting approximation error is embedded into zonotopes. The overall procedure comprises two stages: a prediction phase, in which the prior feasible set is propagated by computing Minkowski sums of zonotopes, and an update phase, in which the predicted zonotope is intersected with a measurement-induced strip and then outer-approximated by a new zonotope. In this way, accurate and robust state estimates are obtained for constrained nonlinear systems, even in situations where classical stochastic filters are difficult to apply due to strong nonlinearities and hard bounds. Furthermore, the same idea can be extended beyond pure state estimation to applications such as fault diagnosis and fault-tolerant control of nonlinear plants. Owing to its flexibility in handling constraints and modeling uncertainties, the proposed approach provides a useful tool for the analysis and control of complex dynamical systems.

4.7 Case Study on Battery Operating Status Analysis

4.7.1 Case 1: Zonotopic Filter

To assess the performance and estimation accuracy of the proposed approach, simulations were performed using an electrothermal-coupled battery model, as described in Section 4.2. The electrical parameters of the ECM adopted in the simulations are listed in Table 4.1.

To extend the discharge duration and emulate the aging process of the power battery, the discharge current was set to 1 A, driving the battery from a fully charged state to complete depletion. This configuration enables the observation of variations in the operating state throughout the entire charge–discharge cycle.

4.7.1.1 State of Charge Simulation

As the number of charge–discharge cycles increases, the internal structure of the battery gradually deteriorates, leading to capacity fade and performance degradation. During this aging process, the rated capacity Q_N continuously decreases, causing deviations in the SOC estimation. Under normal operating conditions, capacity fading progresses rather slowly; therefore, over a limited number of charge–discharge cycles, the resulting SOC estimation error remains modest.

In the simulation study, the initial state and zonotopic parameters are chosen as $x_{c,0} = [5, 5, 1]^{\mathrm{T}}$, $\boldsymbol{p}_{c,0} = [0, 0, 1]^{\mathrm{T}}$, and $G_{c,0} = 0.01I_3$. During the recursive computation, the number of columns of the generator matrix G_c is gradually compressed to six so as to limit the computational burden. The process disturbance and measurement noise satisfy $|w_k|_\infty \leq 10^{-4}$ and $|v_k|_\infty \leq 10^{-3}$, respectively. The sampling period is set to $\Delta T = 5$ s. With these settings, the corresponding system matrices are obtained as follows:

Table 4.1 Parameters of the ECM.

R_0/Ω	R_1/Ω	R_2/Ω	C_1/F	C_2/F	I/A	Q_N/Ah
0.52	0.03	0.04	1531.2	20.21	1	205

$$A_c = \begin{bmatrix} 0.9105 & 0 & 0 \\ 0 & 0.0016 & 0 \\ 0 & 0 & 1 \end{bmatrix}, \quad B_c = \begin{bmatrix} 0.0027 \\ 0.0381 \\ -0.02 \end{bmatrix}, \quad C_c = \begin{bmatrix} -1 & -1 & 0.5158 \end{bmatrix}.$$

Figure 4.6 compares the SOC estimation results obtained using the zonotopic and ellipsoidal filtering methods. The true SOC trajectory is represented by the solid line, while the dashed and dotted lines correspond to the estimation bounds of the zonotope and ellipsoid, respectively. X_{max} and X_{min} denote the upper and lower bounds estimated by the ZF, whereas Z_{max} and Z_{min} represent those obtained from the ellipsoidal approach. It can be observed that the ZF yields tighter and more accurate estimation bounds compared with the ellipsoidal method.

Figure 4.7 illustrates the state estimation results of both filters. Here, U_1 denotes the voltage across the polarization resistance and capacitance corresponding to the first branch, and U_2 represents that of the second branch. The star markers mark the true states, while the solid and dashed lines indicate the estimated states obtained by the zonotopic and ellipsoidal filters, respectively. Because U_1 and U_2 exhibit exponential dynamics, their variations are initially rapid and gradually stabilize as k increases. To provide an intuitive visualization, pairs of the three state variables are plotted in two-dimensional subfigures, illustrating how the zonotopic and ellipsoidal sets enclose the true state trajectory. As shown in Figure 4.7,

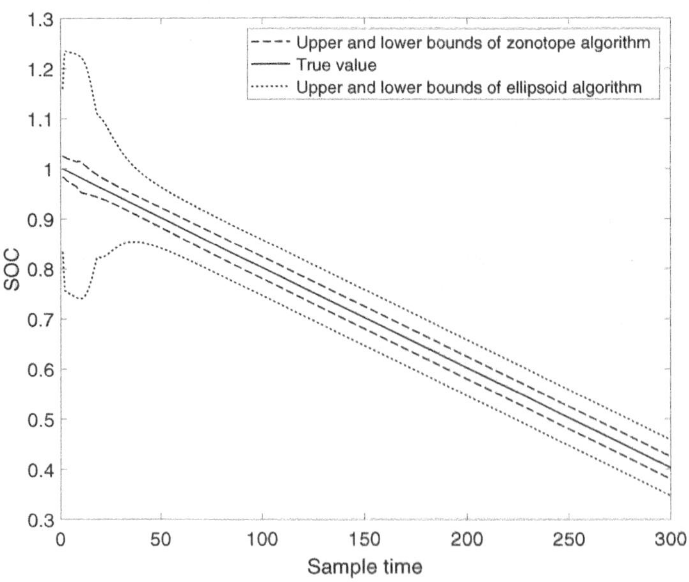

Figure 4.6 Bounds of the SOC estimation.

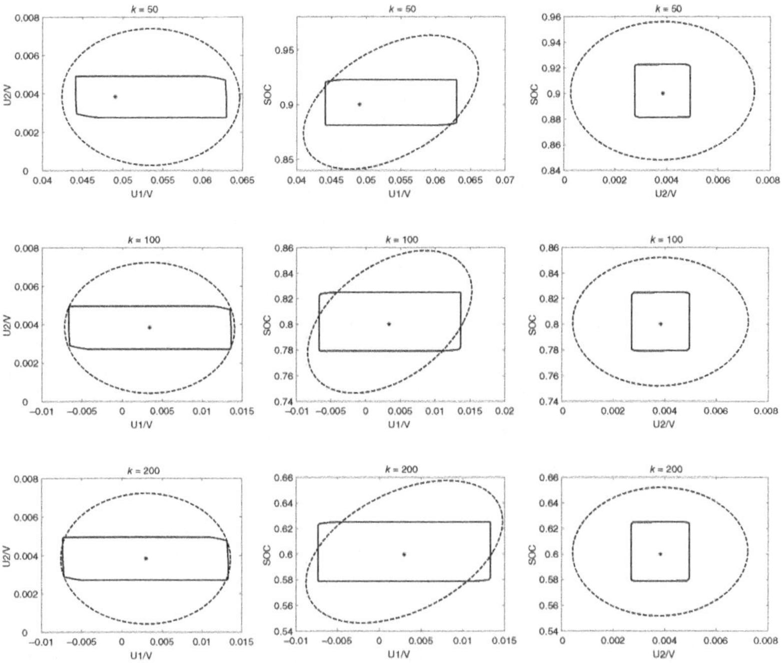

Figure 4.7 Comparison between actual states and estimated states obtained by the two algorithms.

the zonotopic approach achieves a more compact and tighter enclosure of the true state compared with the ellipsoidal method, confirming its superior estimation precision.

4.7.1.2 State of Health Simulation

SOH is adopted as a scalar indicator describing the degradation level of the battery. For a fresh cell, SOH is set to 100% and then decreases gradually as charge–discharge cycling and calendar aging proceed. The loss of SOH is accompanied by observable changes in macroscopic electrical behavior, most prominently an increase in ohmic/internal resistance and a reduction in the maximum usable capacity.

In the proposed model, SOH is treated as a one-dimensional state variable. For the numerical study, the initial state and zonotopic parameters are chosen as $x_{h,0} = 50$, $p_{h,0} = 50$, and $G_{h,0} = 1$. The process disturbance γ_k and measurement noise ε_k are assumed to be UBB, satisfying $|\gamma_k|_\infty \leq 4 \times 10^{-4}$ and $|\varepsilon_k|_\infty \leq 4 \times 10^{-6}$, respectively. The discrete-time SOH dynamics are parameterized by $A_h = 1$ and $C_h = 4 \times 10^{-5}$.

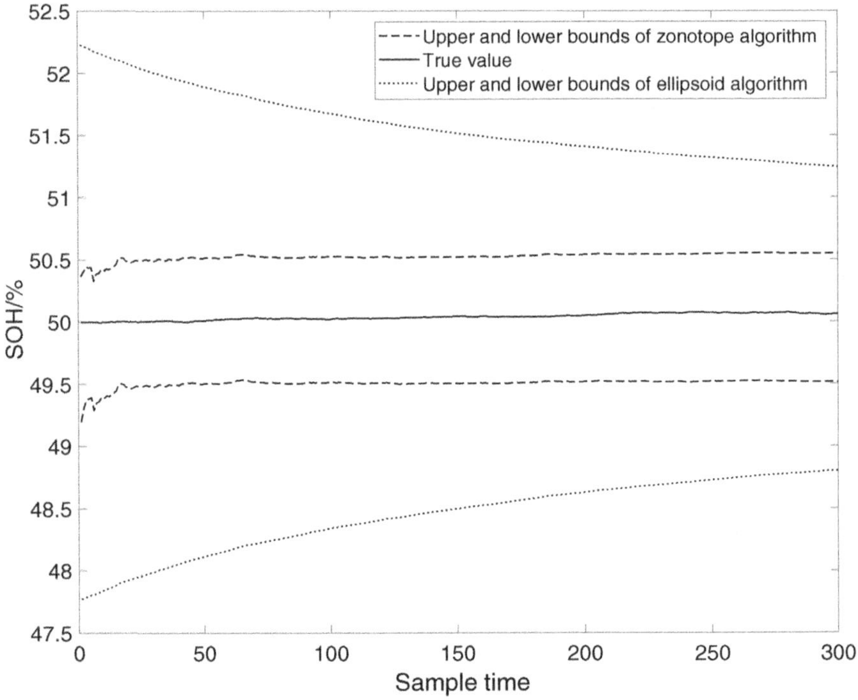

Figure 4.8 Bounds of the SOH estimation.

Figure 4.8 compares the SOH estimation obtained by the zonotopic and ellipsoidal filters. The solid line denotes the true SOH trajectory, while the dashed and dotted lines correspond to the estimation bounds produced by the zonotope- and ellipsoid-based methods, respectively. Specifically, X_{max} and X_{min} denote the upper and lower bounds from the ZF, and Z_{max} and Z_{min} are those from the ellipsoidal filter. As shown in Figure 4.8, the zonotopic approach yields tighter bounds and higher estimation accuracy than the ellipsoidal counterpart.

4.7.2 Case 2: Zonotope and Gaussian Kalman Filters

Li-ion batteries are widely used energy storage units that play essential roles across numerous applications; thus, accurate temperature estimation is of significant importance. In this section, the proposed zonotope and Gaussian Kalman filter (ZGKF) is implemented on the electrothermal Li-ion battery model specified in (4.2) and (4.4). The physical meanings and nominal values of the electrical parameters are reported in Table 4.2, while the corresponding thermal parameters are collected in Table 4.3.

Table 4.2 Electric model parameters.

Parameters	Physical meanings	Values
I	Current	5 A
R_o	Internal resistance	0.0501 Ω
R_1	Electrochemical polarization resistance	0.0298 Ω
R_2	Concentration polarization resistance	0.03819 Ω
C_1	Electrochemical polarization capacity	1789.7 F
C_2	Concentration polarization capacity	5.26 F

Table 4.3 Thermal model parameters.

Parameters	Physical meanings	Values
C_c	Core concentrated heat capacity	$63.5\ \mathrm{J \cdot K^{-1}}$
C_s	Surface concentrated heat capacity	$4.5\ \mathrm{J \cdot K^{-1}}$
R_c	Electrochemical polarization resistance	$1.98\ \mathrm{K \cdot W^{-1}}$
R_u	Concentration polarization resistance	$1.718\ \mathrm{K \cdot W^{-1}}$

By defining the state vector as $x = [T_c,\ T_s]^{\mathrm{T}}$ and the input as $u = [Q_{\text{gen}},\ T_e]^{\mathrm{T}}$, the electrothermal-coupled model can be discretized as:

$$\begin{cases} x_{k+1} = \begin{bmatrix} 1 - \dfrac{\Delta t}{R_c C_c} & \dfrac{\Delta t}{R_c C_c} \\ \dfrac{\Delta t}{R_c C_s} & 1 - \dfrac{\Delta t}{R_c C_s} - \dfrac{\Delta t}{R_u C_s} \end{bmatrix} x_k + \begin{bmatrix} \dfrac{\Delta t}{C_c} & 0 \\ 0 & \dfrac{\Delta t}{R_u C_s} \end{bmatrix} u_k, \\ y_k = \begin{bmatrix} 1 & 0 \\ 0 & 1 \end{bmatrix} x_k. \end{cases} \tag{4.184}$$

Substituting the parameter values into the above expressions and selecting a sampling interval of $\Delta t = 1$ s yields

$$A = \begin{bmatrix} 0.992 & 0.0080 \\ 0.112 & 0.758 \end{bmatrix}, \qquad B = \begin{bmatrix} 0.0157 & 0 \\ 0 & 0.1293 \end{bmatrix}.$$

In the simulation study, the initial state vector is set to $x_0 = [25,\ 25]^{\mathrm{T}}$, and the ambient temperature is fixed at $T_e = 25$ °C. The process noise d_k is modeled as a zero-mean Gaussian sequence, $d_k \sim \mathcal{N}(0, Q)$, with covariance $Q = 0.1^2 I_2$, while the measurement noise e_k follows $e_k \sim \mathcal{N}(0, R)$ with $R = 0.05^2 I_2$. In addition, the UBB disturbances are assumed to satisfy $|\omega_k| \leq [0.01,\ 0.01]^{\mathrm{T}}$ and $|\upsilon_k| \leq$

$[0.01,\ 0.01]^{\mathrm{T}}$, which specify the admissible ranges of the non-stochastic perturbations acting on the system.

Figures 4.9 and 4.10 illustrate the evolution of the estimated core and surface temperatures, along with their corresponding interval bounds. The dash-dotted curves denote the true temperature trajectories. As shown, the ZGKF algorithm yields interval estimates that closely track the true core and surface temperatures. In comparison with the SMKF and EZGKF schemes, the ZGKF yields markedly narrower state-estimation intervals for both T_c and T_s, which reflects a clear improvement in estimation precision. Although the SMKF results approach those of the ZGKF, noticeable deviations remain due to the wrapping effects accumulated when intersecting ellipsoidal sets.

Figure 4.11 compares the estimation intervals at selected sampling instants. The ZGKF intervals, shown by solid boundaries, remain consistently narrower than those obtained by the SMKF and EZGKF algorithms. Overall, the simulation results demonstrate that the proposed ZGKF method provides superior estimation accuracy and stronger convergence properties for electrothermal state estimation in Li-ion batteries.

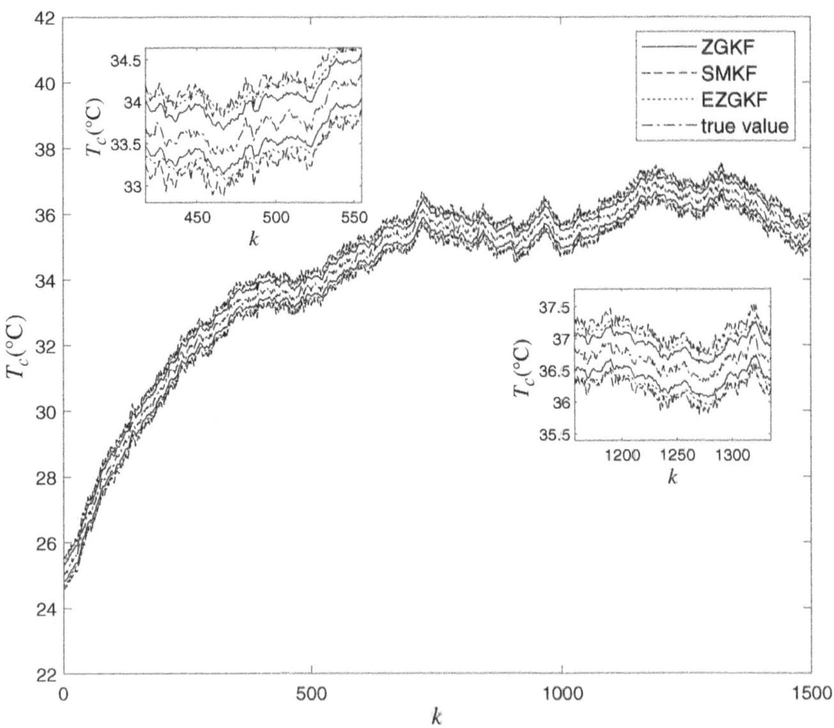

Figure 4.9 Bounds of T_c estimated by different algorithms.

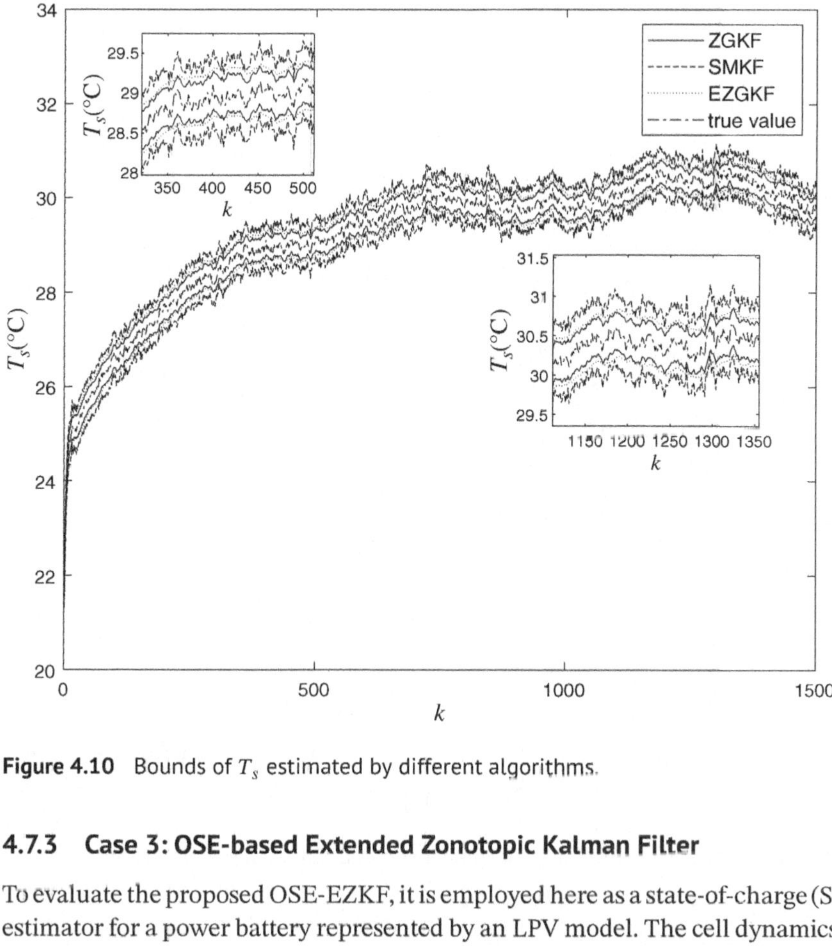

Figure 4.10 Bounds of T_s estimated by different algorithms.

4.7.3 Case 3: OSE-based Extended Zonotopic Kalman Filter

To evaluate the proposed OSE-EZKF, it is employed here as a state-of-charge (SOC) estimator for a power battery represented by an LPV model. The cell dynamics are described by a first-order Thevenin-type equivalent circuit, whose mathematical form is given in (4.1). In this RC network, the electrochemical polarization resistance R_p together with the polarization capacitance C_p captures the polarization behavior of the battery. Since these parameters change with battery temperature and operating conditions along the charge–discharge process, they are modeled as time-varying quantities. The physical interpretation and nominal values of all electrical parameters are reported in Table 4.4.

The mathematical model (4.1) can be rewritten in the transfer-function form

$$G(s) = \frac{U_{oc}(s) - U(s)}{I_c(s)} = R_0 + \frac{R_p}{\tau_p s + 1}, \tag{4.185}$$

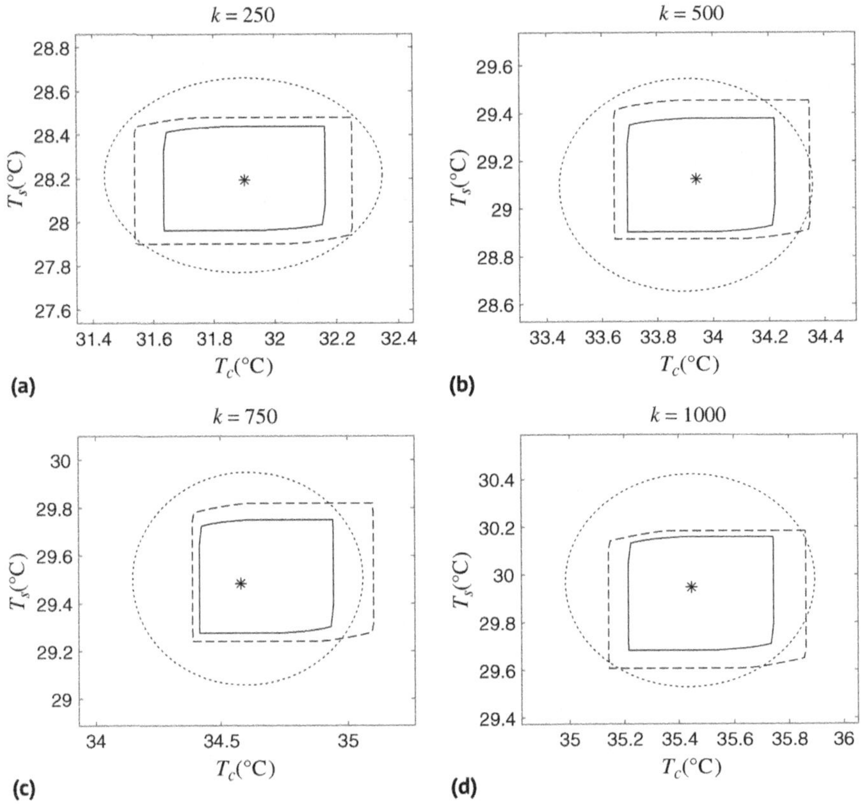

Figure 4.11 State estimation interval comparison at different sampling instants.

Table 4.4 Equivalent model parameters.

Parameter	Physical meaning	Value
I_c	Current	1 A
R_0	Internal resistance	0.52 Ω
$R_p(0)$	Electrochemical polarization resistance	0.03 Ω
$C_p(0)$	Electrochemical polarization capacity	980.3 F
Q_0	Full capacity of the battery	205 A·h
η	Coulomb efficiency	0.6

where R_0 denotes the ohmic resistance and $\tau_p = R_p C_p$ is the time constant. Under constant-current charge/discharge conditions, the first-order backward-difference method leads to the following discrete-time equation:

$$h(k) = \varphi^{\mathrm{T}}(k)\theta(k) + e(k), \tag{4.186}$$

with

$$h(k) = U_{oc}(k) - U(k), \qquad \varphi(k) = \left[h(k-1),\ I_c(k)\right]^{\mathrm{T}},$$

$$\theta(k) = \left[\frac{\tau_p(k)}{\tau_p(k)+T}, \frac{(R_0 + R_p(k))T}{\tau_p(k)+T}\right]^{\mathrm{T}}, \tag{4.187}$$

where $e(k) \in [-0.001,\ 0.001]$ denotes a bounded additive perturbation and $T = 0.5$ s is the sampling interval. With this sampling period, the true time-varying parameters are prescribed as $R_p(k) = 0.03 + 10^{-4}k$ and $C_p(k) = 980.3 + 0.2k$. The identification procedure uses the current and terminal-voltage data recorded during the charge/discharge process to estimate these parameters. Typical identification results are presented in Figure 4.12: the star markers indicate the actual parameter trajectories, whereas the solid rectangles represent the estimated orthotopes. As the iteration index k increases, changes in the slope of the measurement strip lead to a gradual rotation of the orthotope. It can be seen that the

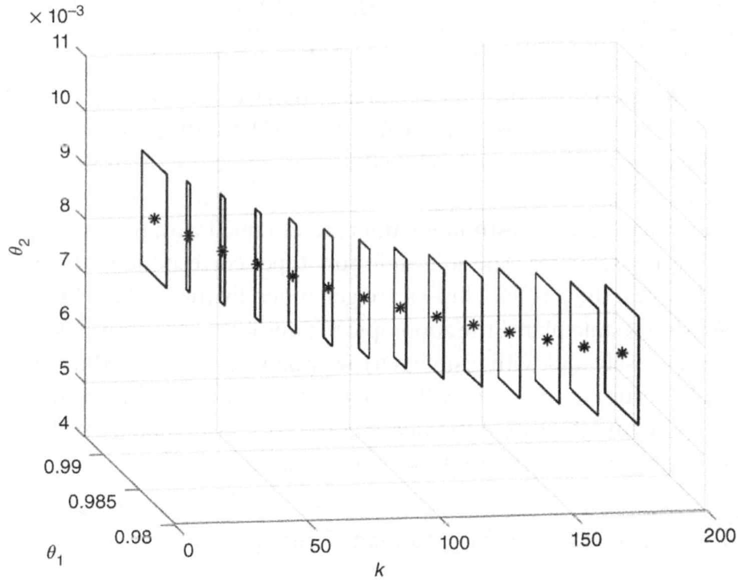

Figure 4.12 Parameter identification of $\theta_1(k)$.

computed orthotopic bounds tightly cover the true parameter curves, demonstrating the capability of the proposed estimator to enclose the real parameter evolution.

From (4.1), the model can be equivalently expressed in the discrete-time state-space form:

$$\begin{cases} \begin{bmatrix} \mathrm{SOC}(k+1) \\ U_p(k+1) \end{bmatrix} = \begin{bmatrix} 1 & 0 \\ 0 & e^{-\frac{T}{\tau_p(k)}} \end{bmatrix} \begin{bmatrix} \mathrm{SOC}(k) \\ U_p(k) \end{bmatrix} + \begin{bmatrix} -\frac{\eta T}{Q_0} \\ (1 - e^{-\frac{T}{\tau_p(k)}}) R_p(k) \end{bmatrix} I_c(k) + w(k), \\ U(k) = U_{oc}(\mathrm{SOC}(k)) - U_p(k) - I_c(k) R_0 + v(k), \end{cases} \tag{4.188}$$

where the open-circuit voltage satisfies

$$U_{oc}(\mathrm{SOC}(k)) = a_{oc}\mathrm{SOC}(k) + b_{oc} = 0.53\,\mathrm{SOC}(k) + 5.58.$$

This model corresponds to (4.96), with time-varying matrices expressed as:

$$A(\theta(k)) = \begin{bmatrix} 1 & 0 \\ 0 & e^{-\frac{T}{\tau_p(\theta(k))}} \end{bmatrix}, \qquad B(\theta(k)) = \begin{bmatrix} -\frac{\eta T}{Q_0} \\ (1 - e^{-\frac{T}{\tau_p(\theta(k))}}) R_p(\theta(k)) \end{bmatrix},$$

with

$$\tau_p(\theta(k)) = \frac{T\theta_1(k)}{1 - \theta_1(k)}, \qquad R_p(\theta(k)) = \frac{\theta_2(k)(\tau_p(k) + T)}{T} - R_0.$$

For the numerical experiment, the state vector is initialized as $x(0) = [1,\ 5]^{\mathrm{T}}$ and the initial generator matrix is chosen as $G(0) = 0.1^2 I_2$. The UBB process disturbance is constrained by $|w(k)|_\infty \leq 10^{-5}$, while the measurement noise is assumed to satisfy $|v(k)|_\infty \leq 10^{-2}$.

Figure 4.13 illustrates the state–estimation intervals obtained with the zonotopic and ellipsoidal filtering schemes. The solid-line zonotopes correspond to the proposed OSE-EZKF, while the dashed ellipses are generated by the EF benchmark. In the two-dimensional state plane, the zonotopic regions follow the true state trajectory much more closely and with visibly narrower bands, indicating that OSE-EZKF attains higher estimation accuracy. Taken together, these simulation results show that the proposed method yields tighter uncertainty bounds and more favorable convergence behavior than the EF-based approach.

4.7.4 Case 4: Constrained Zonotopic Kalman Filtering

To evaluate the performance of the proposed constrained zonotopic Kalman filter (CZKF), we adopt a lithium-ion cell described by an electrothermal coupled model

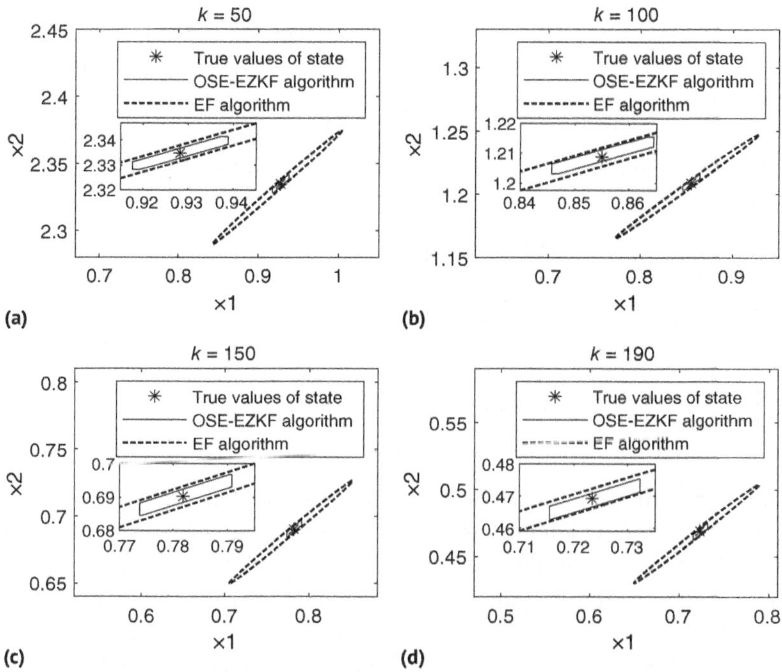

Figure 4.13 Comparison of true and estimated states using the two algorithms.

as the simulation case study and carry out real-time temperature estimation. A Li-ion battery can be viewed as an integrated electrochemical device in which electrical, chemical, and thermal processes interact. The resulting dynamics are highly nonlinear and strongly influenced by the surrounding environment. In addition, the cell temperature is naturally restricted to a safe operating window by physical and safety considerations. These features make state–estimation methods that explicitly handle constraints, such as the CZKF, particularly attractive for battery applications. In this work, the electrothermal coupled model is constructed by combining the second-order Thevenin equivalent circuit with a lumped thermal model [58, 59].

By merging the second-order Thevenin model (4.2) with the thermal dynamics (4.4), the coupled system can be written in the following discrete-time state-space form:

$$
\begin{cases}
x_{k+1} = Ax_k + Bu_k + w_k, \\
y_k = \begin{pmatrix} h(x_{1,k}) \\ x_{2,k} \end{pmatrix} + v_k,
\end{cases}
$$

where $x_k = [T_{c,k}, T_{s,k}]^{\mathrm{T}}$ collects the core and surface temperatures, $u_k = [Q_{\mathrm{gen},k}, T_{e,k}]^{\mathrm{T}}$ contains the heat-generation rate and ambient temperature, and $y_k = [y_{1,k}, y_{2,k}]^{\mathrm{T}}$ represents the measured surface and core temperatures, respectively. The terms ω_k and υ_k denote UBB process and measurement disturbances, and Δt is the sampling interval. The corresponding system matrices are given by

$$A = \begin{bmatrix} 1 - \dfrac{\Delta t}{R_c C_c} & \dfrac{\Delta t}{R_c C_c} \\ \dfrac{\Delta t}{R_c C_s} & 1 - \dfrac{\Delta t}{R_c C_s} - \dfrac{\Delta t}{R_u C_s} \end{bmatrix},$$

$$B = \begin{bmatrix} \dfrac{\Delta t}{C_c} & 0 \\ 0 & \dfrac{\Delta t}{R_u C_s} \end{bmatrix}, \qquad h(x_{1,k}) = 1.1 + 3.4\,e^{-0.08T_{c,k}}.$$

The nominal parameters used in the simulation are listed in Table 4.5.

Under normal operating conditions, the cell is kept in a near–steady thermal state, so the temperature gap between the core and surface remains small and the core temperature varies only slightly. Experimental measurements show that, when the ambient temperature is $T_e = 25$ °C, the core–surface temperature difference is maintained at 5.82 ± 0.05 °C and the core temperature satisfies $T_c \in [35.8\ °\mathrm{C}, 35.9\ °\mathrm{C}]$. On this basis, the state constraint parameters are chosen as:

$$\xi = \begin{bmatrix} 0.05 \\ 0.05 \end{bmatrix}, \qquad D_k = \begin{bmatrix} -5.82 \\ -35.85 \end{bmatrix},$$

$$L_k = \begin{bmatrix} -1 & 1 \\ -1 & 0 \end{bmatrix}.$$

For the simulation study, the initial state is set to $x_0 = [25, 25]^{\mathrm{T}}$, and the initial zonotope is initialized with center $p_0 = [25, 25]^{\mathrm{T}}$ and generator matrix $G_0 = 0.1I_4$. The process and measurement disturbances are assumed to satisfy $|\omega_k|_\infty \leq 0.001$

Table 4.5 Parameters of the electrothermal coupled model.

R_0/Ω	R_1/Ω	R_2/Ω	C_{p1}/F	C_{p2}/F	I/A
0.0501	0.0298	0.03819	1789.7	5.26	5
$C_c/(\mathrm{J}\cdot\mathrm{K}^{-1})$	$C_s/(\mathrm{J}\cdot\mathrm{K}^{-1})$		$R_u/(\mathrm{K}\cdot\mathrm{W}^{-1})$	$R_c/(\mathrm{K}\cdot\mathrm{W}^{-1})$	
63.5	4.5		1.718	1.98	

and $|v_k|_\infty \leq 0.1$, respectively. With a sampling interval of $\Delta t = 1$ s, the corresponding discrete-time system matrices are

$$A = \begin{bmatrix} 0.992 & 0.0080 \\ 0.112 & 0.758 \end{bmatrix}, \qquad B = \begin{bmatrix} 0.0157 & 0 \\ 0 & 0.1293 \end{bmatrix}.$$

To highlight the benefits of the proposed CZKF scheme, its performance is benchmarked against the center-difference zonotopic set-membership filter (CDZSMF) in [60]. The corresponding simulation results are reported in Figures 4.14–4.16.

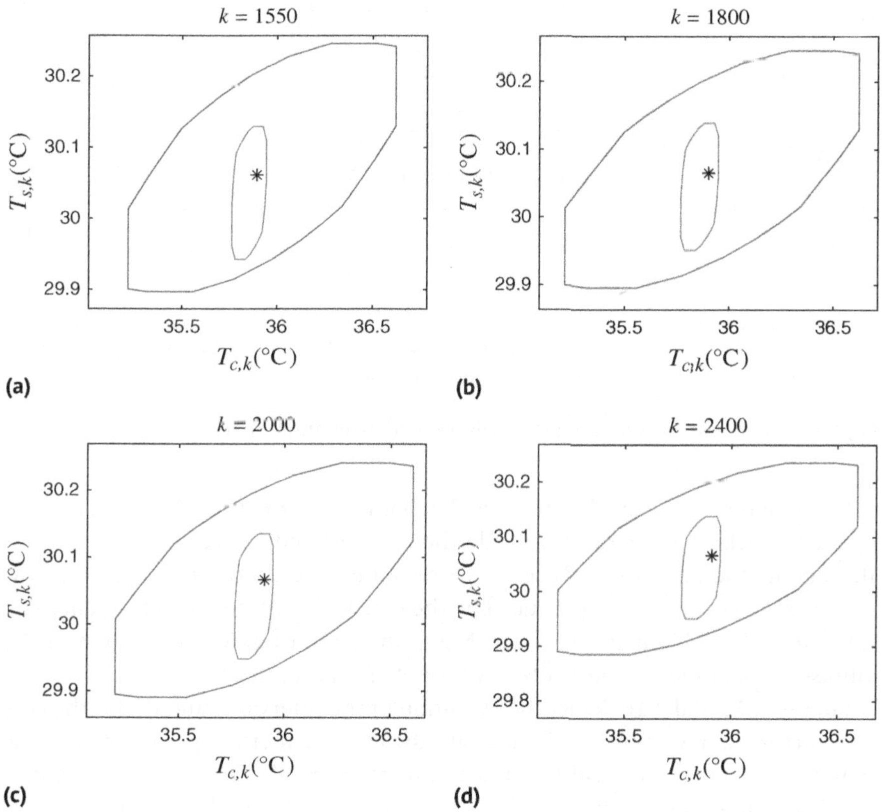

Figure 4.14 Feasible state sets at selected sampling instants.

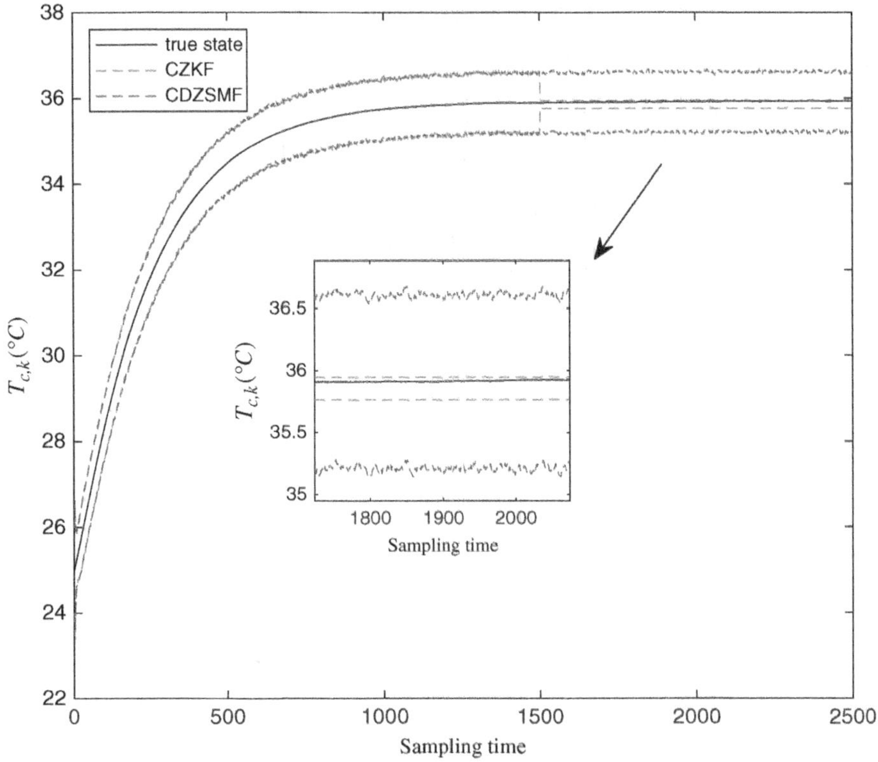

Figure 4.15 Core temperature and corresponding estimation intervals.

From Figure 4.14, it can be observed that both CZKF and CDZSMF keep the true core and surface temperatures inside their corresponding feasible state regions throughout the simulation. Once the system enters the quasi-steady regime, however, the zonotopic region produced by the CZKF shrinks to a much smaller volume than that obtained with CDZSMF, indicating a tighter description of the admissible states and a markedly lower level of conservatism.

Figures 4.15 and 4.16 depict the evolution of the interval estimates for the core and surface temperatures. In both cases, the estimated envelopes follow the real temperature trajectories and continue to cover the actual states. Nevertheless, during steady-state operation the intervals generated by the CZKF are considerably narrower than those of the CDZSMF, highlighting the superior estimation accuracy and improved convergence behavior of the proposed filter.

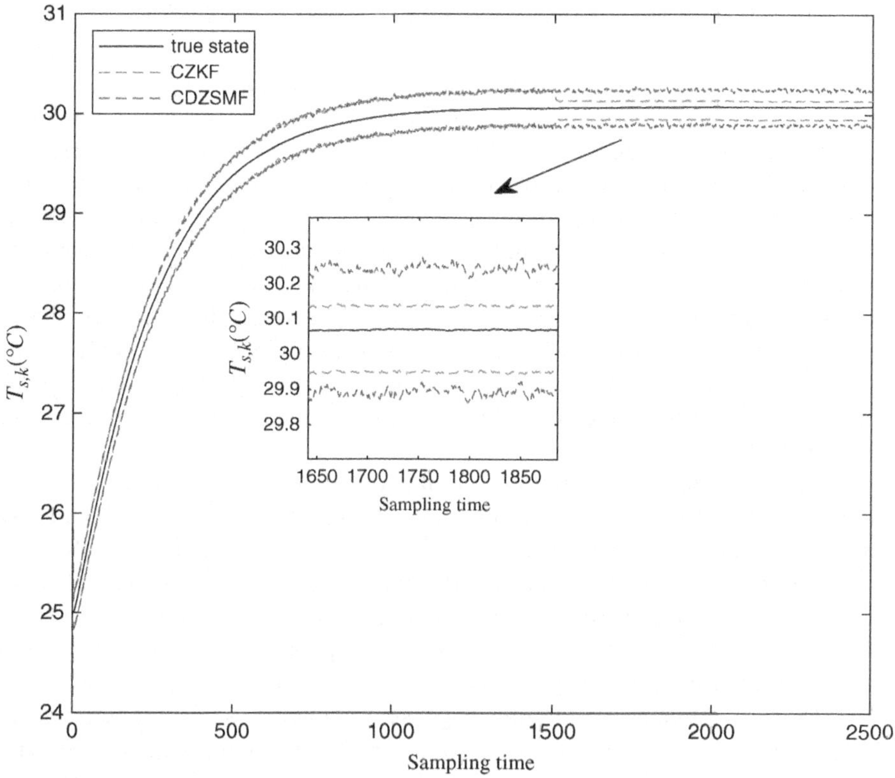

Figure 4.16 Surface temperature and corresponding estimation intervals.

4.8 Concluding Remarks

In this chapter, we introduced a series of advanced state estimation techniques for battery systems, focusing on zonotopic filtering and its various extensions. These algorithms have been designed to tackle the internal states of batteries, such as the SOC and SOH in systems under uncertainty, offering robust solutions for real-time monitoring and control in BMSs. Through the integration of zonotopic filtering with state-of-the-art Kalman filter algorithms, a robust framework has been established for estimating the operational state of batteries under uncertain conditions. By applying these methods, we can effectively handle noise and system uncertainties, ensuring precise estimation of state parameters even in the presence of complex disturbances and nonlinear dynamics.

Several advanced filtering strategies based on the ZF were proposed and evaluated. The ZF provides a straightforward yet efficient approach for state estimation by utilizing zonotopes to bound the uncertainty of the system's state, ensuring that the estimation process is computationally efficient and robust against noise. Building on this, the ZGKF integrate the zonotopic approach with the Gaussian Kalman filter, enabling better handling of linear systems with both process and measurement noise, thus enhancing the accuracy of state estimation. The OSE-EZKF method further extends the zonotopic Kalman filter by introducing orthotope-based expansion to address time-varying parameters, which improves estimation accuracy and robustness. Finally, the CZKF incorporates state constraints into the zonotopic Kalman filter framework, ensuring that the estimates remain within physically feasible boundaries, which is particularly important for applications such as BMSs. These methods collectively provide accurate state estimation, enable real-time monitoring, and ensure proactive maintenance, making them indispensable tools for enhancing the performance and longevity of battery systems.

The research discussed in this chapter lays the groundwork for future advancements in BMSs, where filtering techniques and artificial intelligence (AI) will play an increasingly prominent role. As battery technology continues to evolve, the integration of these methods into smart, adaptive systems will undoubtedly contribute to the advancement of energy storage solutions in various sectors, including electric vehicles and large-scale energy storage systems. In conclusion, the zonotopic filtering techniques covered in this chapter offer a comprehensive solution for state estimation and fault diagnosis in battery systems. They pave the way for future innovations in battery management and contribute to the ongoing improvement of energy storage technologies.

5

Advances in Battery Operating Status Analysis by Filtering and Artificial Intelligence

5.1 Problem Formulation

The state of charge (SOC) is a key indicator for assessing the remaining power of lithium-ion batteries, and its accurate estimation is essential for ensuring the reliability and safety of battery management systems. Since SOC cannot be directly measured, it must be inferred from measurable electrical quantities such as voltage and current. To this end, a practical charge–discharge platform based on 18650 lithium-ion cells was constructed, as shown in Figure 5.1. The platform is built around an 18650 lithium-ion cell, where SOC and OCV are considered the key indicators. A constant-current charging/discharging controller ensures stable operation, producing reliable SOC data. The system includes a real-time monitoring interface to display current, terminal voltage, and SOC variation, while an upper computer records and processes the measured data for subsequent analysis.

In order to establish the mapping between OCV and SOC, a constant-current charge–discharge experiment was conducted at 25 °C. During the test, the battery was periodically discharged in 10% SOC intervals, and after each interval, the cell was rested to record the corresponding OCV value until the cutoff voltage was reached. The resulting OCV–SOC data were then fitted using a polynomial regression, yielding an empirical function

$$U_{\mathrm{OCV}} = f(\mathrm{SOC}), \tag{5.1}$$

which serves as the fundamental relationship for model identification and SOC estimation, providing a direct basis for characterizing the SOC from measurable voltage data.

Building on this static mapping, the overall behavior of the experimental platform can be further described by the first-order Thevenin equivalent circuit model [61], which provides a compact yet effective representation of the battery's electrochemical dynamics. In this model, the state vector is defined by the

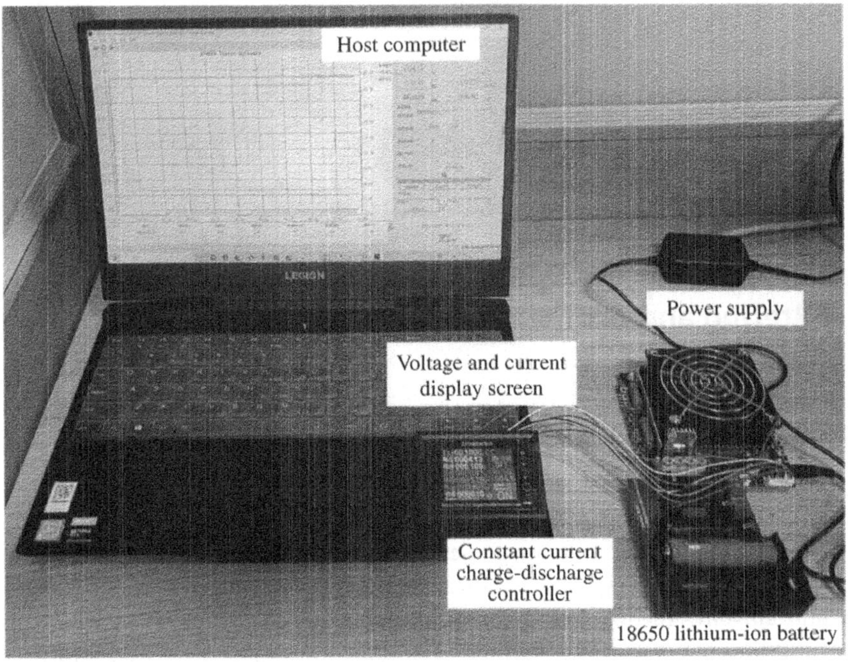

Figure 5.1 Experimental platform for SOC analysis of lithium-ion batteries.

polarization voltage $U_{p,k}$ and the SOC_k, while the terminal voltage U_k serves as the measurement output. The continuous-time model is expressed as:

$$\begin{cases} U_k = U_{\mathrm{OCV}} - I_k R_0 - U_{p,k}, \\ U_{p,k} = e^{-\frac{\Delta t}{R_p C_p}} U_{p,k-1} + \left(1 - e^{-\frac{\Delta t}{R_p C_p}}\right) R_p I_k, \end{cases} \tag{5.2}$$

where R_0 denotes the ohmic resistance, the polarization resistance and capacitance are indicated by R_p and C_p , and $\tau_p = R_p C_p$ defines the polarization time constant. The parameters Q and η represent the nominal capacity and Coulombic efficiency, respectively, while Δt is the sampling interval.

By selecting SOC_k and $U_{p,k}$ as the state variables and U_k as the measurable output, the system is discretized and expressed in the following state-space representation:

$$\begin{cases} \begin{bmatrix} \mathrm{SOC}_{k+1} \\ U_{p,k+1} \end{bmatrix} = \begin{bmatrix} 1 & 0 \\ 0 & e^{-\frac{\Delta t}{R_p C_p}} \end{bmatrix} \begin{bmatrix} \mathrm{SOC}_k \\ U_{p,k} \end{bmatrix} + \begin{bmatrix} -\dfrac{\eta \Delta t}{Q} \\ (1 - e^{-\frac{\Delta t}{R_p C_p}}) R_p \end{bmatrix} I_k + w_k, \\ U_k = U_{\mathrm{OCV},k} - U_{p,k} - I_k R_0 + v_k, \end{cases} \tag{5.3}$$

where $U_{OCV,k}$ denotes the mapping between open-circuit voltage (OCV) and SOC obtained from experimental data. And w_k and v_k denote the process and measurement noise, both assumed to be unknown but bounded (UBB).

Abstracting from the above case, the SOC estimation problem can be embedded into a general discrete-time state-space framework. For linear dynamics with UBB uncertainties, the model takes the form:

$$\begin{cases} x_{k+1} = Ax_k + Bu_k + w_k, \\ y_k = Cx_k + v_k, \end{cases} \tag{5.4}$$

where $x_k \in \mathbb{R}^{n_x}$ denotes the system state, $u_k \in \mathbb{R}^{n_u}$ is the system input, and $y_k \in \mathbb{R}^{n_y}$ represents the output at time step k. Matrices A, B, and C are system parameters with appropriate dimensions. The uncertainties include process noiscs $w_k \in \mathbb{R}^{n_w}$ and measurement errors $v_k \in \mathbb{R}^{n_v}$, both modeled under the UBB assumption.

For generality, the initial state x_0, process disturbances w_k, and measurement noises v_k can be enclosed within zonotopes as $x_0 \in \langle p_0, G_0 \rangle$, $w_k \in \langle 0, W \rangle$, $v_k \in \langle 0, V \rangle$. Here, W and V serve as the corresponding generator matrices.

More generally, for nonlinear dynamics, the system can be expressed as:

$$\begin{cases} x_{k+1} = f(x_k, u_k) + w_k, \\ y_k = h(x_k, u_k) + v_k, \end{cases} \tag{5.5}$$

with $x_k \in \mathbb{R}^n$ denotes the state vector, $y_k \in \mathbb{R}^p$ is the measurement output. Assume that $f(\cdot)$ and $h(\cdot)$ are continuous differentiable nonlinear equations [62], and $f(\cdot)$ and $h(\cdot)$ are both DC functions [63]. This unified abstraction allows the SOC estimation problem to be rigorously studied within the broader framework of discrete-time systems subject to UBB disturbances.

5.2 Introduction to Particle Filter and Particle Swarm Optimization

5.2.1 Particle Filter

5.2.1.1 Problem Setting

For the discrete-time nonlinear state-space model

$$\begin{cases} x_k = f(x_{k-1}) + w_{k-1}, \\ y_k = h(x_k) + v_k, \end{cases} \tag{5.6}$$

where $x_k \in \mathbb{R}^n$ is the state, $y_k \in \mathbb{R}^z$ is the measurement, and w_{k-1}, v_k denote the process and measurement disturbances, respectively. They can be modeled probabilistically (with densities) or specified only through bounds. Given observations $Y_k = \{y_1, \ldots, y_k\}$, filtering seeks the posterior $p(x_k \mid Y_k)$.

5.2.1.2 Bayesian Recursion

Writing the transition and likelihood induced by (5.6) as $p(x_k \mid x_{k-1})$ and $p(y_k \mid x_k)$, the standard predict–update relations are [64]:

$$p(x_k \mid Y^{k-1}) = \int p(x_k \mid x_{k-1})\, p(x_{k-1} \mid Y^{k-1})\, \mathrm{d}x_{k-1}, \tag{5.7}$$

$$p(x_k \mid Y^k) = \frac{p(y_k \mid x_k)\, p(x_k \mid Y^{k-1})}{p(y_k \mid Y^{k-1})}. \tag{5.8}$$

For nonlinear/non-Gaussian cases, the integrals in (5.7) and (5.8) are intractable; PF approximates them by importance sampling.

5.2.1.3 Particle Approximation and Importance Weighting

The posterior is represented by a weighted empirical measure

$$p(x_k \mid Y^k) \approx \sum_{i=1}^{N} w_k^{(i)}\, \delta(x_k - x_k^{(i)}), \tag{5.9}$$

where $\{x_k^{(i)}, w_k^{(i)}\}_{i=1}^{N}$ are particles and normalized weights. Given a proposal (importance) density $q(x_k \mid x_{0:k-1}^{(i)}, Y^k)$, the unnormalized weight recursion is

$$\tilde{w}_k^{(i)} = w_{k-1}^{(i)} \frac{p\left(y_k \mid x_k^{(i)}\right) p\left(x_k^{(i)} \mid x_{k-1}^{(i)}\right)}{q\left(x_k^{(i)} \mid x_{0:k-1}^{(i)}, Y^k\right)}, \qquad w_k^{(i)} = \frac{\tilde{w}_k^{(i)}}{\sum_{j=1}^{N} \tilde{w}_k^{(j)}}. \tag{5.10}$$

A widely used special case is the *bootstrap* PF, which chooses the proposal as the transition prior [65], $q(\cdot) = p(x_k \mid x_{k-1})$. Then

$$\tilde{w}_k^{(i)} = w_{k-1}^{(i)}\, p(y_k \mid x_k^{(i)}). \tag{5.11}$$

Point estimates follow directly from (5.9), e.g., $\hat{x}_k = \sum_{i=1}^{N} w_k^{(i)} x_k^{(i)}$.

5.2.1.4 Degeneracy and Resampling

Weight collapse is monitored by the effective sample size

$$N_{\text{eff}} = \left(\sum_{i=1}^{N} (\bar{w}_k^i)^2 \right)^{-1}. \tag{5.12}$$

When $N_{\text{eff}} < \tau$ (with a user-chosen threshold τ), resampling (e.g., multinomial, systematic, stratified, residual) is performed and resampled particles typically start with uniform weights $1/N$.

5.2.1.5 Canonical PF Recursion

1. Draw particles x_k^i from $q(x_k \mid x_{0:k-1}^i, Y_k)$.
2. Compute weights via (5.10) (or (5.11)) and normalize.
3. If $N_{\text{eff}} < \tau$, resample and continue with uniform weights.
4. Form the estimate, e.g., $\hat{x}_k = \sum_{i=1}^{N} \bar{w}_k^i x_k^i$.

5.2.2 Particle Swarm Optimization

5.2.2.1 Background and Motivation

Particle swarm optimization (PSO) is a gradient-free, population-level search algorithm modeled on the collective dynamics of flocking birds. In PSO, the search space is analogized to a forest, where each particle represents an individual bird searching for food. The optimization process is simulated as particles exploring the search space and updating their positions by sharing information with others, thereby mimicking the cooperative behavior of bird flocks in locating the area with the most abundant food sources. Through this mechanism of social information exchange and self-exploration, PSO enables the swarm to converge efficiently toward the global optimum without relying on gradient information.

5.2.2.2 Algorithmic Formulation

Consider an m-dimensional search space. Initialize a swarm of N particles with velocity $V_i = [v_{i1}, \dots, v_{im}]^{\mathrm{T}} \in \mathbb{R}^m$ and position $X_i = [x_{i1}, \dots, x_{im}]^{\mathrm{T}} \in \mathbb{R}^m$ $(i = 1, \dots, N)$. For each particle, record its individual best position $P_i = [p_{i1}, \dots, p_{im}]^{\mathrm{T}}$, and let $G = [g_1, \dots, g_m]^{\mathrm{T}}$ denote the best position discovered by the swarm. At iteration q, the velocity and position are updated by [66]:

$$V_i^{q+1} = \omega V_i^q + c_1 r_1 \left(P_i^q - X_i^q\right) + c_2 r_2 \left(G^q - X_i^q\right), \tag{5.13}$$

$$X_i^{q+1} = X_i^q + V_i^{q+1}, \tag{5.14}$$

where ω is the inertia weight, c_1 and c_2 are learning factors, and $r_1, r_2 \in (0, 1)$ are uniformly distributed random numbers. The random numbers r_1 and r_2 are resampled at each iteration and may be applied as scalars or componentwise. Whenever $f(X_i^q)$ improves upon $f(P_i^q)$, set $P_i^q \leftarrow X_i^q$; G^q is updated analogously using the best value over all particles. Unless otherwise stated, we use $N = 30$, either a decaying inertia $\omega : 0.9 \rightarrow 0.4$ or the constriction setting $c_1 = c_2 = 2.05$ $(\chi \approx 0.729)$; velocities are clamped by $|v_{i,j}| \leq 0.2\,(U_j - L_j)$, and positions are clipped to $[L, U]$, and reflection is optional.

Equations (5.13) and (5.14) constitute the core PSO iteration used throughout this chapter; additional choices such as velocity clamping, topology, or constriction can be incorporated without changing the basic form.

5.2.2.3 Initialization and Stopping

Initialize positions and velocities componentwise on the search box $[L, U]$:

$$X_i^0 \sim \mathcal{U}([L, U]), \qquad V_i^0 \sim \mathcal{U}([-v_{\max}, v_{\max}]),$$

with $v_{\max} = \alpha\,(U - L)$ and $\alpha \in [0.1, 0.3]$. Set $P_i^0 = X_i^0$ and

$$G^0 = \arg\min_{i=1,\dots,N} f(P_i^0).$$

Terminate when one of the following holds: (i) $q \ge Q_{\max}$; (ii) no improvement, $|f(G^q) - f(G^{q-\tau})| \le \varepsilon$ for τ iterations; (iii) target achieved, $f(G^q) \le f^\star$.

Here $[L, U] \in \mathbb{R}^d$ are the boundary values defining the search interval, and $\mathcal{U}([L, U])$ denotes independent uniform sampling on this box. The velocity cap is $v_{\max} = \alpha\,(U - L)$ (applied per coordinate), with a user-chosen scale $\alpha \in [0.1, 0.3]$. $Q_{\max}$ represents the upper limit on iteration count (positive integer). τ is the patience window for the no–improvement check . $f^\star$ is an optional target objective; if specified, the run stops once $f(G^q) \le f^\star$. Other symbols follow earlier definitions of the swarm: X_i^q, V_i^q, P_i^q, G^q and the particle count N.

5.3 PSO-based Orthometric Hyperparallel Space Filtering

5.3.1 Preliminaries

This subsection establishes several fundamental definitions and properties, and formulates the discrete-time state-space model together with the bounded uncertainty descriptions.

Definition 5.1 Let $p \in \mathbb{R}^n$ be a nonzero vector and $c \in \mathbb{R}$. The (unit-width) strip associated with (p, c) is the set

$$S(p, c) = \{\,\theta \in \mathbb{R}^n : \ |p^\top\theta - c| \le 1\,\}. \tag{5.15}$$

Definition 5.2 [67] The nonempty intersection of n nonparallel n-dimensional strips is referred to as a parallelotope, which can be expressed as:

$$\mathcal{P} = \bigcap_{i=1}^{n} S(p_i, c_i) = \left\{\theta \in \mathbb{R}^n : \ \| P(\theta - \theta_c) \|_\infty \le 1\right\}, \tag{5.16}$$

where $P = [p_1^\top; \dots ; p_n^\top] \in \mathbb{R}^{n\times n}$ is the shape matrix of the parallelepiped $\mathcal{P}$ with P being invertible, and $\theta_c = [c_1, \dots, c_n]^\top \in \mathbb{R}^n$ denotes the center of $\mathcal{P}$.

By letting $T = P^{-1}$, an equivalent representation of the parallelepiped is obtained as:

$$\mathcal{P}(\theta_c, T) = \{\theta : \theta = \theta_c + T\alpha, \ \|\alpha\|_\infty \le 1\}, \tag{5.17}$$

where $T = [t_1, \dots, t_n] \in \mathbb{R}^{n\times n}$ is the generator matrix of the parallelepiped $\mathcal{P}(\theta_c, T)$.

5.3.2 Orthometric Hyperparallel Space Filtering Based on Particle Swarm Optimization

5.3.2.1 Construction of the Orthometric Hyperparallel Particle Search Space

For a linear state-space model in discrete time similar to 5.4, during the prediction phase of the hyperparallel space filter, starting from the hyperparallel space $\mathcal{P}(\theta_{c,k}, T_k)$ that contains the true state at time k, the estimated predicted hyperparallel space at instant $k+1$, expressed as $\mathcal{P}(\hat{\theta}_{c,k+1}, \hat{T}_{k+1})$, is obtained as:

$$\hat{\theta}_{c,k+1} = A\theta_{c,k} + Bu_k, \tag{5.18}$$

$$\hat{T}_{k+1} = AT_k. \tag{5.19}$$

In the update phase, the measurement equation at instant $k+1$ induces the measurement strip S_{k+1}:

$$S_{k+1} = \{\theta : \left|y_{k+1} - C\theta\right| \le V\}, \tag{5.20}$$

where V is the diagonal matrix composed of the maximum magnitudes of the measurement noise. The strip (5.20) is equivalently represented through the intersection of n_y independent strips:

$$\begin{aligned}\bigcap_{a=1}^{n_y} S_{k+1,a} &= \bigcap_{a=1}^{n_y}\{\theta : |C_a\theta - y_{k+1,a}| \le r_{v,a}\} \\ &= \bigcap_{a=1}^{n_y}\left\{\theta : |p_{k+1,a}^{\mathrm{T}}\theta - c_{k+1,a}| \le 1\right\} \\ &= \bigcap_{a=1}^{n_y} S(p_{k+1,a}, c_{k+1,a}),\end{aligned} \tag{5.21}$$

where $p_{k+1,a} = \left(r_{v,a}{}^{-1}C_a\right)^{\mathrm{T}}$, $c_{k+1,a} = r_{v,a}^{-1}y_{k+1,a}$, C_a and $y_{k+1,a}$ denote the a-th components of C and y_{k+1}, respectively, and $r_{v,a}$ is the maximum magnitude of the corresponding measurement noise, $a = 1, 2, \dots, n_y$. Hence, the updated feasible set can be written as the intersection obtained from the predicted hyperparallel space and strips:

$$\mathcal{P}(\hat{\theta}_{c,k+1}, \hat{T}_{k+1}) \bigcap \left(\bigcap_{a=1}^{n_y} S(p_{k+1,a}, c_{k+1,a})\right), \tag{5.22}$$

and $\hat{t}_{k+1,i_p}$ denotes the i_p-th column of the generator matrix $\hat{T}_{k+1}$ ($i_p = 1, 2, \dots, n_x$).

However, since the measurement set S_{k+1} is typically an irregular polytope, a closed-form recursive expression of the intersection in (5.22) is difficult to obtain directly. Lemma 5.1 provides a recursive minimum-volume outer-bounding parallelotope construction to address this issue.

Lemma 5.1 *[68] Assume that the measurement strip $S(p_{k+1,a}, c_{k+1,a})$ intersects the predicted hyperparallel space $\mathcal{P}(\hat{\theta}_{c,k+1}, \hat{T}_{k+1})$. If $p^{\mathrm{T}}_{k+1,a}\hat{t}_{k+1,i_p} < 0$, replace $\hat{t}_{k+1,i_p}$ by $-\hat{t}_{k+1,i_p}$. Then the* minimum-volume *hyperparallel space that outer–bounds the intersection, $\mathcal{P}(\bar{\theta}_{c,k+1}, \bar{T}_{k+1})$, is given by*

$$\mathcal{P}(\bar{\theta}_{c,k+1}, \bar{T}_{k+1}) = \begin{cases} \mathcal{P}(\widetilde{\theta}_{c,k+1}, \widetilde{T}_{k+1}), & \text{if } i^* = 0, \\ \mathcal{P}(\theta^*_{c,k+1}, T^*_{k+1}), & \text{if } i^* \neq 0, \end{cases} \tag{5.23}$$

where

$$i^* = \arg \max_{b=0,\ldots,n_x} \widetilde{p}^{\mathrm{T}}_{k+1,a}\widetilde{t}_{k+1,b}, \tag{5.24}$$

$$\widetilde{t}_{k+1,0} = \widetilde{p}_{k+1,a}/\| \widetilde{p}_{k+1,a}\|^2, \tag{5.25}$$

and

$$\theta^*_{c,k+1} = \widetilde{\theta}_{c,k+1} + \frac{1}{\widetilde{p}^{\mathrm{T}}_{k+1,a}\widetilde{t}_{k+1,i^*}} \widetilde{t}_{k+1,i^*} (\widetilde{c}_{k+1,a} - \widetilde{p}^{\mathrm{T}}_{k+1,a}\widetilde{\theta}_c), \tag{5.26}$$

$$t^*_{k+1,i_p} = \begin{cases} \widetilde{t}_{k+1,i_p} - \dfrac{\widetilde{p}^{\mathrm{T}}_{k+1,a}\widetilde{t}_{k+1,i_p}}{\widetilde{p}^{\mathrm{T}}_{k+1,a}\widetilde{t}_{k+1,i^*}}, & i_p \neq i^*, \\ \dfrac{1}{\widetilde{p}^{\mathrm{T}}_{k+1,a}\widetilde{t}_{k+1,i^*}} \widetilde{t}_{k+1,i^*}, & i_p = i^*, \end{cases} \tag{5.27}$$

$$\widetilde{\theta}_{c,k+1} = \hat{\theta}_{c,k+1} + \sum_{i_p=1}^{n_x} \frac{r^+_{i_p} - r^-_{i_p}}{2} \widehat{t}_{k+1,i_p}, \tag{5.28}$$

$$\widetilde{t}_{k+1,i_p} = \frac{r^+_{i_p} + r^-_{i_p}}{2} \hat{t}_{k+1,i_p}. \tag{5.29}$$

In (5.26) *and* (5.29),

$$\widetilde{p}_{k+1,a} = \frac{2}{r^+_0 + r^-_0} p_{k+1,a}, \tag{5.30}$$

$$\widetilde{c}_{k+1,a} = \frac{2}{r^+_0 + r^-_0}\left(c_{k+1,a} + \frac{r^+_0 - r^-_0}{2}\right), \tag{5.31}$$

$$r_{ip}^{\pm} = \begin{cases} \min\left(1, \dfrac{1 \mp \epsilon_0^{\mp}}{p_{k+1,a}^{\mathrm{T}} \hat{t}_{k+1,i_p}} - 1\right), & p_{k+1,a}^{\mathrm{T}} \hat{t}_{k+1,i_p} \neq 0, \\ 1, & p_{k+1,a}^{\mathrm{T}} \hat{t}_{k+1,i_p} = 0, \end{cases} \tag{5.32}$$

where $r_0^+ = \min\left(1, \epsilon_0^+\right)$ *and* $r_0^- = \min\left(1, -\epsilon_0^-\right)$*, with*

$$\begin{aligned} \epsilon_0^+ &= (p_{k+1,a}^{\mathrm{T}} \hat{\theta}_{c,k+1} - c_{k+1,a}) + \sum_{i=1}^{n_x} p_{k+1,a}^{\mathrm{T}} \hat{t}_{k,i_p}, \\ \epsilon_0^- &= (p_{k+1,a}^{\mathrm{T}} \hat{\theta}_{c,k+1} - c_{k+1,a}) - \sum_{i=1}^{n_x} p_{k+1,a}^{\mathrm{T}} \hat{t}_{k,i_p}. \end{aligned} \tag{5.33}$$

By Lemma 5.1, the computation in (5.22) can be carried out as a sequence of n_y intersections between a strip and a hyperparallel space: after each iteration, the resulting hyperparallel space is intersected with the next strip; after n_y such iterations, the final minimum-volume hyperparallel space that contains the true state at instant $k+1$ is denoted by $\mathcal{P}(\theta_{c,k+1}, T_{k+1})$.

Since the particle swarm must be initialized within a hyperparallel space but the coordinate-wise ranges inside a general parallelotope geometry are not explicit, the set $\mathcal{P}(\theta_{c,k+1}, T_{k+1})$ cannot be used directly as the PSO search domain. A practical remedy is to construct an *orthometric* hyperparallel space that shares the same lower and upper bounds in every dimension; its orthogonality renders the coordinate ranges explicit for particle initialization and subsequent iterations.

Theorem 5.1 *The orthometric hyperparallel space* $\mathcal{P}(\theta_{c,k+1}^o, T_{k+1}^o)$ *that shares the same lower and upper bounds with* $\mathcal{P}(\theta_{c,k+1}, T_{k+1})$ *can be expressed as:*

$$\hat{\theta}_{c,k+1}^o = \begin{bmatrix} \dfrac{\mu_{1,k+1}^{\max} + \mu_{1,k+1}^{\min}}{2} \\ \dfrac{\mu_{2,k+1}^{\max} + \mu_{2,k+1}^{\min}}{2} \\ \vdots \\ \dfrac{\mu_{n_x,k+1}^{\max} + \mu_{n_x,k+1}^{\min}}{2} \end{bmatrix}, \tag{5.34}$$

$$\hat{T}_{k+1}^o = \mathrm{diag}\left\{\frac{\mu_{1,k+1}^{\max} - \mu_{1,k+1}^{\min}}{2}, \frac{\mu_{2,k+1}^{\max} - \mu_{2,k+1}^{\min}}{2}, \dots, \frac{\mu_{n_x,k+1}^{\max} - \mu_{n_x,k+1}^{\min}}{2}\right\}, \tag{5.35}$$

where $\mu_{d,k+1}^{\max}$ *and* $\mu_{d,k+1}^{\min}$ *denote the extremal bounds of* $\mathcal{P}(\theta_{c,k+1}, T_{k+1})$ *in the d-th dimension, for* $d = 1, 2, \dots, n_x$.

Proof 5.1. According to Definition 5.2, the hyperparallel space $\mathcal{P}(\theta_{c,k+1}, T_{k+1})$ that contains the true state at time $k+1$ admits the generator representation

$$\mathcal{P}(\theta_{c,k+1}, T_{k+1}) = \{\theta : \theta = \theta_{c,k+1} + T_{k+1}\alpha,\ \|\alpha\|_\infty \leq 1\}. \tag{5.36}$$

Let t_{i_p} denote the i_p-th column vector of the generator matrix T_{k+1}, where $i_p = 1, 2, \dots, n_x$. Then the generator matrix in (5.35) can be expressed as:

$$T_{k+1} = [t_1, t_2, \dots, t_{n_x}] \in \mathbb{R}^{n_x \times n_x}. \tag{5.37}$$

Accordingly, each vertex z_{i_z} of the hyperparallel space can be written as:

$$z_{i_z} = \theta_{c,k+1} + \sum_{i_p=1}^{n_x} t_{i_p}\alpha_{i_p} \in \mathbb{R}^{n_x}, \tag{5.38}$$

where $i_z = 1, 2, \dots, 2^{n_x}$ and $\alpha_{i_p} \in \{-1, 1\}$. Collecting all vertices at instant $k+1$ gives:

$$Z_{k+1} = [z_{1,k+1}, z_{2,k+1}, \dots, z_{2^{n_x},k+1}] \in \mathbb{R}^{n_x \times 2^{n_x}}. \tag{5.39}$$

The extremal values of the d-th row of Z_{k+1} yield the bounds of $\mathcal{P}(\theta_{c,k+1}, T_{k+1})$ in the d-th dimension:

$$\begin{cases} \mu_{d,k+1}^{\max} = \max\left(Z_{k+1}(d, i_z)\right), \\ \mu_{d,k+1}^{\min} = \min\left(Z_{k+1}(d, i_z)\right), \end{cases} \tag{5.40}$$

where $Z_{k+1}(d, i_z)$ denotes the iz-th element in the d-th row of Z_{k+1} ($d = 1, 2, \dots, n_x$).

Based on the extrema in (5.40), the center $\hat{\theta}^o_{c,k+1}$ and generator matrix $\hat{T}^o_{k+1}$ of $\mathcal{P}(\hat{\theta}^o_{c,k+1}, \hat{T}^o_{k+1})$ can be obtained as:

$$\hat{\theta}^o_{c,k+1} = \begin{bmatrix} \dfrac{\mu_{1,k+1}^{\max} + \mu_{1,k+1}^{\min}}{2} \\ \dfrac{\mu_{2,k+1}^{\max} + \mu_{2,k+1}^{\min}}{2} \\ \vdots \\ \dfrac{\mu_{n_x,k+1}^{\max} + \mu_{n_x,k+1}^{\min}}{2} \end{bmatrix},$$

$$\hat{T}^o_{k+1} = \operatorname{diag}\left\{\frac{\mu_{1,k+1}^{\max} - \mu_{1,k+1}^{\min}}{2}, \frac{\mu_{2,k+1}^{\max} - \mu_{2,k+1}^{\min}}{2}, \dots, \frac{\mu_{n_x,k+1}^{\max} - \mu_{n_x,k+1}^{\min}}{2}\right\},$$

and each vertex $z^o_{i_z}$ of the orthometric hyperparallel space satisfies

$$
\begin{aligned}
z^o_{i_z} &= \hat{\theta}^o_{c,k+1} + \sum_{i_p=1}^{n_x} t^o_{i_p} \alpha_{i_p} \\
&= \begin{bmatrix} \dfrac{\mu^{\max}_{1,k+1} + \mu^{\min}_{1,k+1}}{2} \\ \dfrac{\mu^{\max}_{2,k+1} + \mu^{\min}_{2,k+1}}{2} \\ \vdots \\ \dfrac{\mu^{\max}_{n_x,k+1} + \mu^{\min}_{n_x,k+1}}{2} \end{bmatrix} + \begin{bmatrix} \dfrac{\mu^{\max}_{1,k+1} - \mu^{\min}_{1,k+1}}{2} \\ 0 \\ \vdots \\ 0 \end{bmatrix} \alpha_{i_p} \\
&+ \begin{bmatrix} 0 \\ \dfrac{\mu^{\max}_{2,k+1} - \mu^{\min}_{2,k+1}}{2} \\ \vdots \\ 0 \end{bmatrix} \alpha_{i_p} + \cdots + \begin{bmatrix} 0 \\ 0 \\ \vdots \\ \dfrac{\mu^{\max}_{n_x,k+1} - \mu^{\min}_{n_x,k+1}}{2} \end{bmatrix} \alpha_{i_p}
\end{aligned} \tag{5.41}
$$

where $t^o_{i_p}$ denotes the i_p-th column of $\hat{T}^o_{k+1}$. Consequently, the upper and lower bounds of $\mathcal{P}(\hat{\theta}^o_{c,k+1}, \hat{T}^o_{k+1})$ are

$$
\begin{cases}
z^{o,\max}_{i_z} = \begin{bmatrix} \mu^{\max}_{1,k+1} \\ \mu^{\max}_{2,k+1} \\ \vdots \\ \mu^{\max}_{n_x,k+1} \end{bmatrix}, & \alpha_{i_p} = 1, \\
z^{o,\min}_{i_z} = \begin{bmatrix} \mu^{\min}_{1,k+1} \\ \mu^{\min}_{2,k+1} \\ \vdots \\ \mu^{\min}_{n_x,k+1} \end{bmatrix}, & \alpha_{i_p} = -1.
\end{cases} \tag{5.42}
$$

Therefore, from (5.40) and (5.42), the constructed orthometric hyperparallel space $\mathcal{P}(\hat{\theta}^o_{c,k+1}, \hat{T}^o_{k+1})$ and the original hyperparallel space $\mathcal{P}(\theta_{c,k+1}, T_{k+1})$ share identical lower and upper bounds in every dimension.

Based on Theorem 5.1, space $\mathcal{P}(\hat{\theta}^o_{c,k+1}, \hat{T}^o_{k+1})$ possesses the boundary values as $\mathcal{P}(\theta_{c,k+1}, T_{k+1})$. Since each dimension of the orthometric hyperparallel space can be regarded as an interval, the coordinate range of every particle X_i within the swarm, during both random initialization and subsequent iterative optimization, can be described as:

$$\mu_{d,k+1}^{\min} \leq x_{i,d} \leq \mu_{d,k+1}^{\max}, \tag{5.43}$$

where $x_{i,d}$ denotes the d-th component of particle X_i.

5.3.2.2 PSO Update Based on Orthometric Hyperparallel Space

After obtaining $\mathcal{P}(\hat{\theta}^o_{c,k+1}, \hat{T}^o_{k+1})$ and using it as the particle search space, there is a need to establish a new fitness function to evaluate the quality of the particles. Meanwhile, since the particle swarm after iterations typically forms an irregular region, it cannot be directly employed as the new state space for recursive filtering. Therefore, we derive the updated orthometric hyperparallel space during the filtering iterations.

Theorem 5.2 *Suppose that the particles* $\{X_{i,0}^{k+1}\}_{i=1}^{n}$ *are randomly initialized at time* $k+1$ *within the orthometric hyperparallel space* $\mathcal{P}(\hat{\theta}^o_{c,k+1}, \hat{T}^o_{k+1})$ *that contains the true state. After iterative optimization, the orthometric hyperparallel space that encloses the updated swarm* $\{X_i^{k+1}\}_{i=1}^{n}$ *could be formulated as:*

$$\begin{aligned} \{X_i^{k+1}\}_{i=1}^{n} &\in \mathcal{P}(\overline{\theta}^o_{c,k+1}, \overline{T}^o_{k+1}) \\ &= \left\{\theta : \theta = \overline{\theta}^o_{c,k+1} + \mathrm{diag}(\overline{T}^o_{k+1})\,\alpha,\ \|\alpha\|_\infty \leq 1\right\}, \end{aligned} \tag{5.44}$$

where

$$\overline{\theta}^{o,d}_{c,k+1} = \frac{\max\ e_d^{\mathrm{T}}X + \min\ e_d^{\mathrm{T}}X}{2},$$

$$\text{s.t.}\ X \in \{X_{i,k+1}\}_{i=1}^{n}, \tag{5.45}$$

$$\overline{T}^{o,d}_{k+1} = \frac{\max\ e_d^{\mathrm{T}}X - \min\ e_d^{\mathrm{T}}X}{2},$$

$$\text{s.t.}\ X \in \{X_{i,k+1}\}_{i=1}^{n}, \tag{5.46}$$

with e_d *denoting the d-th column of the identity matrix* I_{n_x}*, for* $d = 1, 2, \ldots, n_x$.

Proof 5.2. In order for the particles to eventually scatter around the true state, a fitness function must be defined to obtain both the individual best and global best values used for updating particle positions. At the l-th iteration, each particle's location corresponds to a potential estimate of the system state at the present instant. The predicted measurement associated with each particle can be written as:

$$\hat{y}^{i,l}_{k+1} = CX^{l}_{i,k+1} + \hat{v}_k, \tag{5.47}$$

where $\hat{\upsilon}_k$ is a random variable selected from the hyperparallel space $\mathcal{P}(0, V)$ and serves as an estimate of the system noise. Since the true state is unknown and cannot be directly employed for particle evaluation, the measured output—which contains information about the true state—is incorporated to form the fitness function at $k+1$ instant:

$$\text{fitness}(k+1) = \exp[-(y_{k+1} - \hat{y}^{i,l}_{k+1})^2], \tag{5.48}$$

where y_{k+1} refers to the true measurement at $k+1$ time, while $\hat{y}^{i,l}_{k+1}$ indicates the predicted measurement of i-th particle.

Each particle's velocity and location are then updated from iteration l to $l+1$ as:

$$V^{l+1}_{i,k+1} = \omega V^{l}_{i,k+1} + c_1 r_1 \left(\text{Pbest}^{l}_{k+1} - X^{l}_{i,k+1}\right) + c_2 r_2 \left(\text{Gbest}^{l}_{k+1} - X^{l}_{i,k+1}\right), \tag{5.49}$$

$$X^{l+1}_{i,k+1} = X^{l}_{i,k+1} + V^{l}_{i,k+1}, \tag{5.50}$$

where $Pbest^{l}_{k+1}$ and $Gbest^{l}_{k+1}$ denote the individual and global best positions, respectively.

If the i-th particle exceeds the boundary of exploration space in dimension d, that is,

$$X^{l+1}_{i,k+1} \notin \mathcal{P}\left(\hat{\theta}^{o}_{c,k+1}, \hat{T}^{o}_{k+1}\right), \tag{5.51}$$

its position must be updated accordingly:

1. If $x^{d,l+1}_{i,k+1} > \mu^{\max}_{d,k+1}$,

$$x^{d,l+1}_{i,k+1,\text{new}} = x^{d,l}_{i,k+1} + (\mu^{\max}_{d,k+1} - x^{d,l}_{i,k+1}) \times \text{rand},$$

2. If $x^{d,l+1}_{i,k+1} < \mu^{\min}_{d,k+1}$,

$$x^{d,l+1}_{i,k+1,\text{new}} = x^{d,l}_{i,k+1} + (\mu^{\min}_{d,k+1} - x^{d,l}_{i,k+1}) \times \text{rand},$$

where $x^{d,l+1}_{i,k+1}$ denotes the d-th component of particle $X^{l+1}_{i,k+1}$, $x^{d,l+1}_{i,k+1,\text{new}}$ is the corrected value, $\mu^{\max}_{d,k+1}$ and $\mu^{\min}_{d,k+1}$ define the bounding values of the d-th dimension, $d = 1, 2, \dots, n_x$, and rand denotes a uniformly sampled random variable over $[0, 1]$.

After ensuring that each particle remains inside $\mathcal{P}(\hat{\theta}^{o}_{c,k+1}, \hat{T}^{o}_{k+1})$, the fitness values of the updated particles are recomputed, and the individual and global bests are updated accordingly. The iterations proceed until the swarm's optimal fitness value meets the stopping threshold δ, at which point the resulting particle set

$$A(k+1) = \{X_{1,k+1}, X_{2,k+1}, \dots, X_{n,k+1}\}, \tag{5.52}$$

is considered to be distributed around the true state.

However, the region of high probability formed by the scattered particles is irregular, and a traditional hyperparallel space cannot easily enclose all of them.

To overcome this limitation, a new orthometric hyperparallel space enclosing the entire swarm is constructed by solving $2n_x$ linear programs to determine its bounds:

$$\beta_{k+1}^{d} = \max\ e_d^{\mathrm{T}} X, \quad \text{s.t. } X \in A(k+1), \tag{5.53}$$

$$\beta_{k+1}^{d+n_x} = \min\ e_d^{\mathrm{T}} X, \quad \text{s.t. } X \in A(k+1), \tag{5.54}$$

where β_{k+1}^{d} and $\beta_{k+1}^{d+n_x}$ represent the extremal bounds of the d-th state estimate, and e_d is the d-th column of the identity matrix. The enclosing orthometric hyperparallel space is then given by

$$\mathcal{P}\left(\bar{\theta}_{c,k+1}^{o}, \bar{T}_{k+1}^{o}\right) = \left\{\theta : \theta = \bar{\theta}_{c,k+1}^{o} + \mathrm{diag}(\bar{T}_{k+1}^{o})\,\alpha, \ \ \|\alpha\|_{\infty} \leq 1\right\} \tag{5.55}$$

with center and generator defined as:

$$\bar{\theta}_{c,k+1}^{o,d} = \frac{\beta_{k+1}^{d} + \beta_{k+1}^{d+n_x}}{2}, \tag{5.56}$$

$$\bar{T}_{k+1}^{o,d} = \frac{\beta_{k+1}^{d} - \beta_{k+1}^{d+n_x}}{2}. \tag{5.57}$$

Figure 5.2, taking the two-dimensional case for demonstration, illustrates the process of updating the orthometric hyperparallel space at instant $k+1$ using PSO. Within the space $\mathcal{P}(\hat{\theta}_{c,k+1}^{o}, \hat{T}_{k+1}^{o})$, the swarm is first randomly initialized. Based on the fitness function fitness, both the individual best and global best particles are selected, which guide the swarm to move toward the optimal particle. As a result, after iterations, the final swarm is distributed around the true state. The updated orthometric hyperparallel space is then obtained according to the distribution of the swarm.

In summary, the proposed PSO-based orthometric hyperparallel space filtering (PSO–OHSF) algorithm proceeds as follows:

Step 1: Set $k = 1$, initialize the hyperparallel space $\mathcal{P}(\theta_{c,1}, T_1)$, and define the total sampling horizon K;

Step 2: Use (5.23) to derive the intersection formed by the predicted parallelotope set and the strips, yielding $\mathcal{P}(\theta_{c,k}, T_k)$ that contains the true state;

Step 3: Apply (5.34) and (5.35) to construct the orthometric hyperparallel set $\mathcal{P}(\hat{\theta}_{c,k}^{o}, \hat{T}_k^{o})$ that shares the same bounds as $\mathcal{P}(\theta_{c,k}, T_k)$, and use it as the search space for swarm initialization and optimization;

Step 4: Initialize n particles within $\mathcal{P}(\hat{\theta}_{c,k}^{o}, \hat{T}_k^{o})$, compute their fitness values using (5.48), select the individual and global best particles, and then update particle velocity and position using (5.49) and (5.50), obtaining a swarm $\{X_i^k\}_{i=1}^{n}$ distributed around the true state;

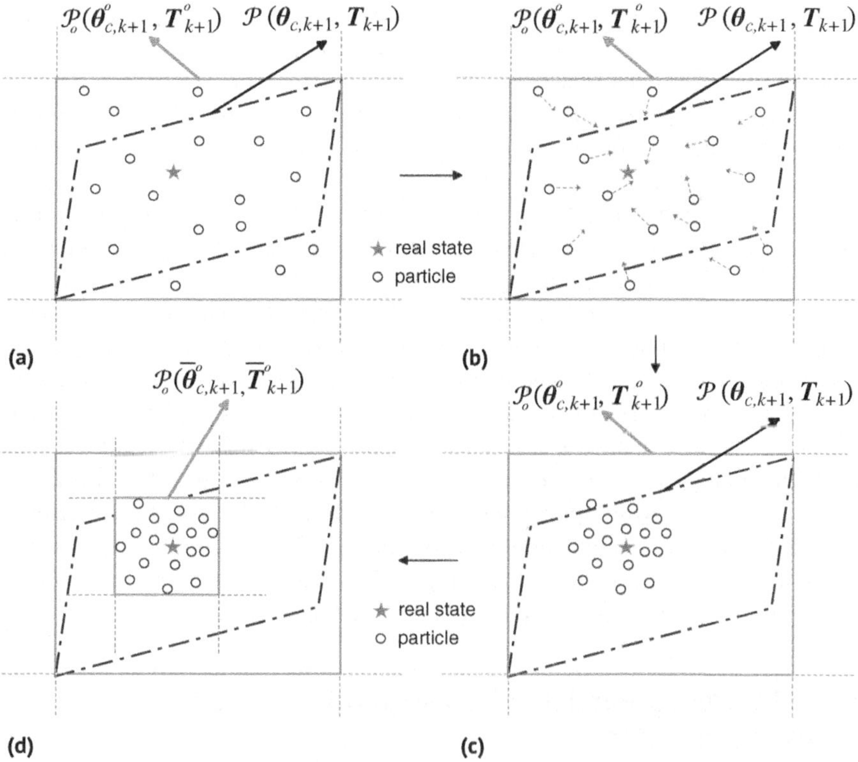

Figure 5.2 Process of particle swarm updating the orthometric hyperparallel space. (a) Initial particle generation, (b) particle iterative optimization, (c) particle optimization finished, and (d) orthogonal hyperparallel space after construction and update.

Step 5: Perform linear programming on $\{X_i^k\}_{i=1}^n$ using (5.53), (5.54), and compute the space $\mathcal{P}(\bar{\theta}_{c,k}^o, \bar{T}_k^o)$ that encloses all particles using (5.56) and (5.57);

Step 6: Update $k = k + 1$ and repeat from Step 2. Once $k = K$, the iteration terminates, and the extremal bounds together with the orthometric hyperparallel space's center are output as the final state estimation results.

5.3.3 Conclusion

In this subsection, a PSO–OHSF algorithm is developed to address the state estimation problem under uniformly bounded but unknown (UBB) noise. The key concept is to construct an orthometric hyperparallel space that shares the same bounds as the feasible hyperparallel space derived from system dynamics and measurement constraints, thereby offering a geometrically regular domain that guides the initialization and evolution of the particle swarm. A fitness function, rigorously formulated from system observations and model residuals, is employed to evaluate the likelihood of each particle representing the true system state. Through iterative position and velocity updates, the swarm is adaptively steered toward regions with higher probability of containing the true state, ensuring convergence while preventing premature stagnation.

To account for the irregular spatial distribution of optimized particles, a refined orthometric hyperparallel space is reconstructed via a linear programming procedure, which circumscribes the optimized particles within a compact region and defines a reduced feasible domain with minimal conservatism. Consequently, the method realizes an equilibrium between conservative stability and estimation precision: the compactness of hyperparallel spaces ensures bounded-error resilience, while the adaptiveness of PSO enhances the tracking precision of the true state.

Beyond the linear system framework considered in this study, the proposed PSO–OHSF approach provides a generalizable foundation for robust estimation in broader engineering applications, particularly in scenarios characterized by bounded uncertainties and nonlinear or time-varying dynamics, such as fault-tolerant control, battery state monitoring, and networked system security, thereby highlighting its versatility and potential for future extensions.

5.4 Improved Particle Filter Algorithm Based on the Parallelotope

5.4.1 Preliminaries

For clarity, essential notations and properties are provided as follows. The notations used in this paper are explained below: $\mathbb{R}^n$ denotes n-dimensional real spaces. $\det(A)$ refers to the determinant of a matrix A. α represents a vector, and the matrix $\mathrm{diag}\{\alpha\}$ indicates a diagonal matrix with the entries of α along its diagonal. $|\cdot|$ signifies the absolute value, $\|\cdot\|_2$ refers to the Euclidean norm and $\|\cdot\|_\infty$ to the infinite norm. $\mathcal{A} \bigcap \mathcal{B}$ represents the intersection between $\mathcal{A}$ and $\mathcal{B}$.

Definition 5.3 [69] Given a real constant c and a nonzero vector p, an n-dimensional strip is defined as:

$$S(p, c) = \{\theta : |p^{\mathrm{T}}\theta - c| \leq 1\}, \tag{5.58}$$

where $p \in \mathbb{R}^n$ is a normal vector of the hyperplanes that bound the strip and $c \in \mathbb{R}$ indicates their displacement from the origin, θ represents the point that satisfies $\theta \in S(p, c)$ in Eq. (5.58).

Definition 5.4 [70] An n-order zonotope arises from the nonempty intersection of $m \geq n$ strips, which can be defined as:

$$\mathcal{Z}(\theta_c, T) = \{\theta : \theta = \theta_c + T\alpha, \parallel \alpha \parallel_\infty \leq 1\}, \tag{5.59}$$

with $T \in \mathbb{R}^{n\times m}$ the generator matrix and $\theta_c \in \mathbb{R}^n$ its center. Specially, when $m = n$, the resulting polytope reduces to a parallelotope that can be represented in its generator as:

$$\mathcal{P}(\theta_c, T) = \{\theta : \theta = \theta_c + T\alpha, \|\alpha\|_\infty \leq 1\}, \tag{5.60}$$

where $T \in \mathbb{R}^{n\times n}$ and $\theta_c \in \mathbb{R}^n$.

Definition 5.5 [63] The function f is the difference of convex (DC) function, and there are two convex functions g_1 and g_2 that satisfies

$$f(x) = g_1(x) - g_2(x). \tag{5.61}$$

Definition 5.6 The symbol $C_m^n (n < m)$ is given by:

$$C_m^n = \frac{m(m-1)(m-2)\cdots(m-n+1)}{n(n-1)(n-2)\cdots 1}. \tag{5.62}$$

Property 4 [70] The Minkowski sum of zonotope $\mathcal{Z}(\theta_1, T_1)$ and $\mathcal{Z}(\theta_2, T_2)$ yields another zonotope, expressed as:

$$\mathcal{Z}(\theta_1, T_1) \oplus \mathcal{Z}(\theta_2, T_2) = \mathcal{Z}(\theta_1 + \theta_2, [T_1\ T_2]). \tag{5.63}$$

In particular, when the generator matrices T_1 and T_2 of the two parallelotopes are diagonal matrices with same dimensions, Eq. (5.63) can be expressed as:

$$\mathcal{P}(\theta_1, T_1) \oplus \mathcal{P}(\theta_2, T_2) = \mathcal{P}(\theta_1 + \theta_2, T_1 + T_2). \tag{5.64}$$

Property 5 [63] If f is a DC function, then every part $f_i (i = 1, 2, \cdots, n)$ of the $f = [f_1, f_2, \cdots, f_n]^{\mathrm{T}}$ is a DC function.

5.4.2 Design of the Improved Particle Filter Based on the Parallelotope

5.4.2.1 Prediction

For the nonlinear system described in Eq. (5.5), N initial particles of the state are sampled from initial-state probability density function $p(x_0)$, denoted as $\{x_0^{i,+}\}_{i=1}^N$. The prior particles $x_k^{i,-}$ are then obtained by

$$x_k^{i,-} = f(x_{k-1}^{i,+}) + w_{k-1}^i, \quad (i = 1, \cdots, N). \tag{5.65}$$

The objective of the prediction step is to construct a parallelotope that encloses all feasible prior particles, achieved through the parallelotope set-membership filtering algorithm.

When applying the parallelotope algorithm to nonlinear systems, the procedure resembles that of the extended Kalman filter (EKF). Specifically, this model is first linearized, after which the linear set-membership filtering algorithm—comprising prediction and update steps—is directly applied to the linearized system. Linearizing $\hat{x}_{k-1}$ yields the following approximation:

$$\begin{aligned} f(x_{k-1}) &= f(\hat{x}_{k-1}) + A_k(x_{k-1} - \hat{x}_{k-1}) + e_{k-1} \\ &= f_L(x_{k-1}) + e_{k-1}, \end{aligned} \tag{5.66}$$

where A_k denotes the Jacobian matrix of the nonlinear function $f = [f_1, f_2, \cdots, f_n]^{\mathrm{T}}$, and e_{k-1} represents the linearization error of the state equation. For the function f, each component f_i must be considered individually in order to determine the corresponding error interval e_{k-1}. For illustrative purposes, only a single component f_i is explicitly derived.

According to Property 5 and Definition 5.5, the two convex functions $g_{1,i}$ and $g_{2,i}$ take the expressions:

$$g_{1,i}(x) = f_i(x) + \alpha x^{\mathrm{T}} x, \qquad g_{2,i}(x) = \alpha x^{\mathrm{T}} x, \tag{5.67}$$

and by linearizing $g_{1,i}$ and $g_{2,i}$ at the current state estimate $\hat{x}_{k-1}$, one obtains

$$\overline{g}_{1,i}(x_{k-1}) = g_{1,i}(\hat{x}_{k-1}) + \upsilon_1^{\mathrm{T}}(x_{k-1} - \hat{x}_{k-1}) \leq g_{1,i}(x_{k-1}), \tag{5.68}$$

$$\overline{g}_{2,i}(x_{k-1}) = g_{2,i}(\hat{x}_{k-1}) + \upsilon_2^{\mathrm{T}}(x_{k-1} - \hat{x}_{k-1}) \leq g_{2,i}(x_{k-1}), \tag{5.69}$$

where υ_1 and υ_2 denote the subgradients of g_1 and g_2, respectively, which can be computed as:

$$\upsilon_1 = \frac{\partial g_{1,i}(\hat{x}_{k-1})}{\partial x}, \qquad \upsilon_2 = \frac{\partial g_{2,i}(\hat{x}_{k-1})}{\partial x}. \tag{5.70}$$

Referring to Eq. (5.66), the linearization error e_{k-1}^i can be bounded by DC programming, leading to

$$e_{k-1,\min}^i = \min_{\hat{x}_{k-1} \in V_{\mathcal{P}}} \{\overline{g}_{1,i}(x_{k-1}) - g_{2,i}(x_{k-1}) - f_{i,L}(x_{k-1})\}, \tag{5.71}$$

$$e_{k-1,\max}^i = \max_{\hat{x}_{k-1} \in V_{\mathcal{P}}} \{g_{1,i}(x_{k-1}) - \overline{g}_{2,i}(x_{k-1}) - f_{i,L}(x_{k-1})\}, \tag{5.72}$$

with $V_{\mathcal{P}}$ representing the hyperparallel domain of the state at $k-1$. After evaluating f_i, the error interval is given as:

$$\begin{aligned} e_{k-1} &= [e_{k-1,\min}, e_{k-1,\max}] \\ &= \left\{e_{k-1} \in \mathbb{R}^n \,\middle|\, e_{k-1,\min}^i \leq e_{k-1}^i \leq e_{k-1,\max}^i\right\}, \\ &\quad (i = 1, \cdots, n), \end{aligned} \tag{5.73}$$

which can further be reformulated into the parallelotope representation

$$\mathcal{P}(\theta_{e_{k-1}}, \text{diag}\{r_{e_{k-1}}\}) = \{\theta : \theta = \theta_{e_{k-1}} + \text{diag}\{r_{e_{k-1}}\}\alpha,\ \|\alpha\|_\infty \leq 1\}, \tag{5.74}$$

where $\theta_{e_{k-1}} = (e_{k-1,\min} + e_{k-1,\max})/2 \in \mathbb{R}^n$, $r_{e_{k-1}} = (e_{k-1,\max} - e_{k-1,\min})/2$, and $\text{diag}\{r_{e_{k-1}}\} \in \mathbb{R}^{n\times n}$.

According to the nonlinear system in Eq. (5.5) and its linearized form in Eq. (5.66), at k, the system state takes the form

$$x_k = f_L(x_{k-1}) + e_{k-1} + w_{k-1}. \tag{5.75}$$

This expression indicates that the prediction of the system state is composed of three components: (i) the linearization function f_L, which yields the hyperparallel space $\mathcal{P}(f(\hat{x}_{k-1}), A_k T_{k-1})$; (ii) the hyperparallel space associated with the linearization error, $\mathcal{P}(\theta_{e_{k-1}}, \text{diag}\{r_{e_{k-1}}\})$; and (iii) the hyperparallel space corresponding to the process noise, $\mathcal{P}(0, \text{diag}\{r_w\})$. By applying the Minkowski sum operation together with Property 4, one obtains

$$\begin{aligned}\mathcal{Z}(\hat{x}_{k,k-1}, T_{k,k-1}) = &\ \mathcal{P}(f(\hat{x}_{k-1}), A_k T_{k-1}) \oplus \mathcal{P}(\theta_{e_{k-1}}, \text{diag}\{r_{e_{k-1}}\}) \oplus \mathcal{P}(0, \text{diag}\{r_w\}) \\ = &\ \mathcal{Z}\left(f(\hat{x}_{k-1}) + \theta_{e_{k-1}}, \left[A_k T_{k-1} \quad \text{diag}\{r_{e_{k-1}} + r_w\}\right]\right).\end{aligned} \tag{5.76}$$

Lemma 5.2 *[71] For the zonotope $\mathcal{Z}(\theta_c, T)$, the half-space representation of its associated polytope is given by*

$$Hx \leq k, \quad H = \begin{bmatrix} H^+ \\ -H^+ \end{bmatrix}, \quad k = \begin{bmatrix} k^+ \\ k^- \end{bmatrix}. \tag{5.77}$$

where

$$H_l^+ = \frac{\Theta(\mathbb{T})^{\mathrm{T}}}{\| \Theta(\mathbb{T}) \|_2}, \quad l = 1, \cdots, C_m^{n-1}, \tag{5.78}$$

$$\Theta(T) = [\cdots, (-1)^{h+1} \det(T^h), \cdots]^{\mathrm{T}}, \quad h = 1, \cdots, n, \tag{5.79}$$

$$k_l^+ = H_l^+ \theta_c + \Delta k_l, \quad k_l^- = -H_l^+ \theta_c + \Delta k_l, \tag{5.80}$$

$$\Delta k_l = \sum_{v=1}^{m} \left| H_l^+ t_v \right|, \tag{5.81}$$

and $\mathbb{T}$ denotes a matrix consisting of $m - n + 1$ vectors arbitrarily selected from T. Here, H_l^+ is the l-th row of H^+, and k_l is the l-th element of k.

Because the computation of the zonotope in the update step is intricate, the minimal-volume parallelotope enclosing $\mathcal{Z}(\hat{x}_{k,k-1}, T_{k,k-1})$ can be determined using Theorem 5.3.

Theorem 5.3 *Given the zonotope $\mathcal{Z}(\theta_c, T)$, the minimal-volume parallelotope that wraps $\mathcal{Z}(\theta_c, T)$ can be expressed as:*

$$\mathcal{P}(\theta_c, \overline{G}) = \left\{\theta : \theta = \theta_c + \overline{G}\alpha, \| \alpha \|_\infty \leq 1\right\}, \tag{5.82}$$

where

$$\overline{G} = ([\overline{\varphi}_1, \cdots, \overline{\varphi}_n]^{\mathrm{T}})^{-1}, \tag{5.83}$$

$$\overline{\varphi}_i = H^+_{vm_i} / \Delta k_{vm_i}. \tag{5.84}$$

Proof 5.3. According to Lemma 5.2, the zonotope $\mathcal{Z}(\theta_c, T)$ with $\theta_c \in \mathbb{R}^n$ and $T \in \mathbb{R}^{n\times m}$ can be reformulated as the half-space representation of its corresponding polytome

$$\{x \in \mathbb{R}^n : Hx \leq k\}$$

This polytope is bounded by $2C_m^{n-1}$ half-spaces. For each half-space $\{H_l^+ x \leq k_l^+\}$, there exists a parallel half-space $\{-H_l^+ x \leq k_l^-\}$. Together, these two parallel half-spaces define a strip $S(\varphi_l, d_l)$, expressed as:

$$\begin{aligned} S(\varphi_l, d_l) &= \{x : -k_l^- \leq H_l^+ x \leq k_l^+\} \\ &= \{x : H_l^+\theta_c - \Delta k_l \leq H_l^+ x \leq H_l^+\theta_c + \Delta k_l\} \\ &= \{x : |H_l^+ x - H_l^+\theta_c| \leq \Delta k_l\} \\ &= \{x : |\varphi_l^{\mathrm{T}} x - d_l| \leq 1\}, \quad l = 1, \cdots, u, \end{aligned} \tag{5.85}$$

where $\varphi_l^{\mathrm{T}} = H_l^+ / \Delta k_l$, $d_l = H_l^+\theta_c / \Delta k_l$, and $u = C_m^{n-1}$.

By Definition 5.4, the zonotope $\mathcal{Z}$ admits the representation as the intersection of u strips:

$$\mathcal{Z} = \bigcap_{l=1}^{u} S(\varphi_l, d_l). \tag{5.86}$$

Selecting any n strips from the set of u strips, the intersection of these n strips yields a parallelotope $\mathcal{P}$ that encloses $\mathcal{Z}$:

$$\begin{aligned} \mathcal{Z} \subset \mathcal{P} &= \bigcap_{l=1}^{n} S(\varphi_l, d_l) \\ &= \{x : \|F_j(x - \theta_c)\|_\infty \leq 1\}, \quad j = 1, \cdots, C_u^n, \end{aligned} \tag{5.87}$$

where F_j is a matrix composed of n rows φ_l^{T} selected arbitrarily from the columns of $\varphi = [\varphi_1, \cdots, \varphi_u]^{\mathrm{T}}$, such that

$$F_j = [\varphi_{g1(q)}, \cdots, \varphi_{gn(q)}]_j^{\mathrm{T}} \in \mathbb{R}^{n\times n}.$$

Here, $\varphi^{\mathrm{T}}_{gi(q)}$ corresponds to different possible selections of n columns from φ, with $1 \leq gi(q) \leq u$, $i = 1, \cdots, n$, and $q = 1, \cdots, u$.

From Definition 5.4, the strips generated by $\mathcal{Z}$ form a nonempty intersection. Consequently, any selection of n distinct strips also produces a nonempty intersection, implying that F_j is invertible.

To compute the volume of the parallelotope, one must first determine its generation matrix G_j of the parallelotope. Substituting $G_j = F_j^{-1}$, we get:

$$\mathcal{P}(\theta_c, G_j) = \{x : x = \theta_c + G_j\alpha, \| \alpha \|_\infty \leq 1\}. \tag{5.88}$$

The generation matrix with the smallest parallelotope volume is denoted as:

$$G^* = \arg \min_{j=1,\cdots,C_u^n} 2^n |\det(G_j)| \tag{5.89}$$

and we can obtain the minimal-volume wrap parallelotope in its generator representation as:

$$\mathcal{P}(\theta_c, G^*) = \{x : x = \theta_c + G^*\alpha, \| \alpha \|_\infty \leq 1\}, \tag{5.90}$$

where

$$G^* = \overline{F^*}^{-1} = ([\overline{\varphi}_1, \cdots, \overline{\varphi}_n]^{\mathrm{T}})^{-1}. \tag{5.91}$$

Suppose the φ_l^{T} selected for the minimal volume of the parallelotope that wraps $\mathcal{Z}(\theta_c, T)$ are $\varphi_{vm_1}^{\mathrm{T}}, \cdots, \varphi_{vm_n}^{\mathrm{T}}$, we can obtain:

$$[\overline{\varphi}_1, \cdots, \overline{\varphi}_n]^{\mathrm{T}} = [\varphi_{vm_1}, \cdots, \varphi_{vm_n}]^{\mathrm{T}}, \tag{5.92}$$

$$\overline{\varphi}_i = H_{vm_i}^{+} / \Delta k_{vm_i}, \tag{5.93}$$

Figure 5.3 illustrates the procedure for selecting the bounding parallelotope of minimal volume in a two-dimensional state space. For instance, when $n = 2$ and $m = 4$, we obtain $u = C_m^{n-1} = 4$ and $\max(j) = C_u^n = 6$. This implies that $\mathcal{Z}$ is determined by four strips $\{S(\varphi_l, d_l)\}_{l=1}^{4}$, which can be combined to construct a parallelotope enclosing $\mathcal{Z}$. By applying Theorem 5.3, the minimal bounding parallelotope is derived.

According to Theorem 5.3, when setting $G^*_{k,k-1} = \hat{G}_{k,k-1}$, the minimal-volume parallelotope that encloses $\mathcal{Z}(\hat{x}_{k,k-1}, T_{k,k-1})$ is expressed as:

$$\begin{aligned}\mathcal{P}(\hat{x}_{k,k-1}, \hat{G}_{k,k-1}) = \{x_{k,k-1} : x_{k,k-1} = \hat{x}_{k,k-1} + \hat{G}_{k,k-1}\alpha,\\ \|\alpha\|_\infty \leq 1\},\end{aligned} \tag{5.94}$$

where $\mathcal{P}(\hat{x}_{k,k-1}, \hat{G}_{k,k-1})$ represents the feasible region of the state space that contains all admissible prior particles generated in the prediction step.

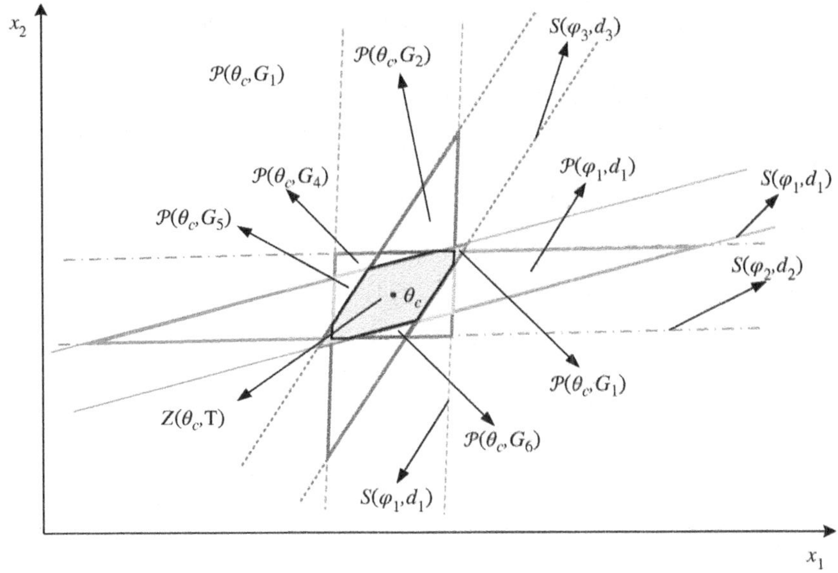

Figure 5.3 Selection of the minimal bounding parallelotope in the two-dimensional state space.

5.4.2.2 Update

During this phase, the feasible domain of particles must be further contracted by incorporating the measurement set. To construct this set, the measurement equation is linearized and reformulated via DC programming in a manner analogous to the state prediction step. After linearization, the measurement equation takes the form:

$$h(x_k) = h(\hat{x}_{k,k-1}) + B_k(x_k - \hat{x}_{k,k-1}) + o_k, \tag{5.95}$$

where B_k represents the Jacobian matrix of the function h, and o_k represents the linearization error. Following Eqs. (5.67)–(5.73), the bounding parallelotope of the linearization error o_k is given by $\mathcal{P}(\tau_{o_k}, \text{diag}\{r_{o_k}\})$. Subsequently, by evaluating the Minkowski sum of o_k with the measurement noise υ_k, we introduce an auxiliary variable $\nu = o_k + \upsilon_k$. Consequently, we have

$$\begin{aligned} \nu &\in \mathcal{P}(\tau_{o_k}, \text{diag}\{r_{o_k}\}) \oplus \mathcal{P}(0, \text{diag}\{r_\upsilon\}) \\ &= \mathcal{P}(\tau_{o_k}, \text{diag}\{r_{o_k} + r_\upsilon\}), \end{aligned} \tag{5.96}$$

where r_v denotes the measurement noise bound as introduced in Eq. (5.5). Therefore, at time step k, the measurement set is formulated as:

$$S_k = \Big\{x : \big|B_k x_k - (y_k - h(\hat{x}_{k,k-1}) + B_k \hat{x}_{k,k-1} - \tau_{o_k})\big| \\ \leq \mathrm{diag}\{r_{o_k} + r_{v_k}\}\Big\}. \tag{5.97}$$

Equation (5.97) admits an equivalent representation as the intersection of z independent strips:

$$S_k = \bigcap_{i=1}^{z} S_{k,i} = \bigcap_{i=1}^{z} \Big\{x : \big|B_{k,i} x - y_{k,i}^a\big| \leq r_i^a\Big\}, \tag{5.98}$$

where $y_{k,i}^a$ and r_i^a $(i = 1, \cdots, z)$ are the components of $y_k - h(\hat{x}_{k,k-1}) + B_k \hat{x}_{k,k-1} - \theta_{o_k}$ and $\mathrm{diag}\{r_{o_k} + r_{v_k}\}$, respectively. Finally, according to Definition 5.3, the measurement set can be normalized as:

$$S_k = \bigcap_{i=1}^{z} S_{k,i} = \bigcap_{i=1}^{z} \Big\{x : \big|p_{k,i}^{\mathrm{T}} x - c_{k,i}\big| \leq 1\Big\}. \tag{5.99}$$

The intersection $\mathcal{P}(\hat{x}_{k,k-1}, \hat{G}_{k,k-1}) \cap S_k$ generally results in a complex and irregular polytope, which is not suitable for iterative computation within the algorithm. To simplify the procedure, this work adopts a sequential strategy in which only one strip $S_{k,i}$ is intersected with the parallelotope at a time. At each step, the minimal bounding parallelotope is obtained, and the process is repeated sequentially until all strips $S_{k,1}, \cdots, S_{k,z}$ have been considered.

As an illustrative example, consider $\mathcal{P}(\hat{x}_{k,k-1}, \hat{G}_{k,k-1}) \cap S(c_{k,1}, p_{k,1})$. Following an approach similar to the recursive optimal hyperparallel outbounding (ROPO) algorithm in [72], the initial parameters are set as:

$$\theta_{c,k} = \hat{x}_{k,k-1}, \quad T_k = [t_{k,1}, \cdots, t_{k,n}] = \hat{G}_{k,k-1}, \quad p_0 = p_{k,1}, \quad c_0 = c_{k,1}.$$

If $p_0^{\mathrm{T}} t_{k,i} < 0$, then $t_{k,i}$ is replaced with $-t_{k,i}$ until the condition $p_0^{\mathrm{T}} t_{k,i} \geq 0$ is satisfied. Two scalars are then defined as:

$$\begin{cases} \epsilon_0^+ = (p_0^{\mathrm{T}} \theta_{c,k} - c_0) + \sum\limits_{i=1}^{n} p_0^{\mathrm{T}} t_{k,i}, \\ \epsilon_0^- = (p_0^{\mathrm{T}} \theta_{c,k} - c_0) - \sum\limits_{i=1}^{n} p_0^{\mathrm{T}} t_{k,i}. \end{cases} \tag{5.100}$$

To enforce a tighter overlap between the measurement strip and the parallelotope, the strip is contracted as:

$$\overline{p}_0 = \frac{2}{r_0^+ + r_0^-} p_0, \quad \overline{c}_0 = \frac{2}{r_0^+ + r_0^-}\left(c_0 + \frac{r_0^+ - r_0^-}{2}\right), \tag{5.101}$$

where $r_0^+ = \min(1, \epsilon_0^+)$ and $r_0^- = \min(1, -\epsilon_0^-)$.

Subsequently, by scaling the vectors $t_{k,i}$, the non-overlapping region between $S(c_0, p_0)$ and $\mathcal{P}(\theta_{c,k}, T_k)$ is reduced. The resulting contracted parallelotope is defined for $i = 1, \cdots, n$ as:

$$\begin{cases} \bar{\theta}_c = \theta_{c,k} + \sum_{i=1}^{n} \frac{r_i^+ - r_i^-}{2} t_{k,i}, \\ \bar{t}_i = \frac{r_i^+ + r_i^-}{2} t_{k,i}, \end{cases} \tag{5.102}$$

with

$$r_i^\pm = \begin{cases} \min\left(1, \frac{1 \mp \epsilon_0^\mp}{p_0^\mathrm{T} t_{k,i}} - 1\right), & \text{if } p_0^\mathrm{T} t_{k,i} \neq 0, \\ 1, & \text{if } p_0^\mathrm{T} t_{k,i} = 0. \end{cases} \tag{5.103}$$

Since the intersection $\mathcal{P}(\overline{\theta}_c, \overline{T}) \cap S(\overline{c}_0, \overline{p}_0)$ is characterized by $n + 1$ strips, there exist $n + 1$ candidate parallelotopes enclosing this intersection. The one with minimal volume is determined by

$$i^* = \arg \max_{j=0,\cdots,n} \overline{p}_0^\mathrm{T} \overline{t}_j, \quad \bar{t}_0 = \frac{\overline{p}_0}{|\overline{p}_0|^2}. \tag{5.104}$$

The minimal volume out bounding parallelotope can be obtained by

$$\mathcal{P}_k(\hat{\theta}_k, \hat{T}_k) = \begin{cases} \mathcal{P}(\bar{\theta}_c, \bar{T}), & \text{if } i^* = 0, \\ \mathcal{P}(\theta_c^*, T^*), & \text{otherwise}, \end{cases} \tag{5.105}$$

where

$$\begin{aligned} \theta_c^* &= \bar{\theta}_c + \frac{1}{\bar{p}_0^\mathrm{T} \bar{t}_{i^*}} \bar{t}_{i^*} (\bar{c}_0 - \bar{p}_0^\mathrm{T} \bar{\theta}_c), \\ t_i^* &= \begin{cases} \bar{t}_i - \frac{\bar{p}_0^\mathrm{T} \bar{t}_i}{\bar{p}_0^\mathrm{T} \bar{t}_{i^*}}, & \text{if } i \neq i^*, \\ \frac{1}{\bar{p}_0^\mathrm{T} \bar{t}_{i^*}} \bar{t}_{i^*}, & \text{otherwise}. \end{cases} \end{aligned} \tag{5.106}$$

After repeating the above algorithm z times, the minimal outbounding parallelotope $\mathcal{P}(\overline{x}_{k,k}, \overline{G}_{k,k})$ can be obtained, where $\overline{x}_{k,k} = \hat{\theta}_k$ and $\overline{G}_{k,k} = \hat{T}_k$.

After obtaining $\mathcal{P}(\overline{x}_{k,k}, \overline{G}_{k,k})$, the feasible space of the particles is the state space limited by the $\mathcal{P}(\overline{x}_{k,k}, \overline{G}_{k,k})$, and then the preservation and abandonment of prior particles can be completed by Theorem 5.4.

Theorem 5.4 *Assume that the set of prior particles generated at time k is* $X_k^- = [x_k^{1,-}, x_k^{2,-}, \cdots, x_k^{N,-}]$, *and that the feasible particle domain is referred to as* $\mathcal{P}(\overline{x}_{k,k}, \overline{G}_{k,k})$. *The weight of particles lying outside this feasible domain is assigned as:*

$$w_k^i = 0, \quad if\ P_k(x_k^{i,-} - \hat{x}_{k,k}) > 1, \tag{5.107}$$

where $P_k = \overline{G}_{k,k}^{-1}$. *The retained particles are denoted as* $x_{k,r}^{i,-}$, *while those discarded are denoted as* $x_{k,d}^{i,-}$.

To preserve the total number of particles N, new effective particles $x_{k,\text{new}}^{i,-}$ *are generated through the solution of the following optimization problem:*

$$\min_{x_{k,\text{new}}^{i,-} \in \mathcal{P}(\bar{x}_{k,k}, \bar{G}_{k,k})} -\left(\lg\left[p_{x_k^e}(x_{k,\text{new}}^{i,-} - x_k^{i,-})\right] + \lg\left[p_{\upsilon_k}(y_k - h(x_{k,\text{new}}^{i,-}))\right]\right), \tag{5.108}$$

where x_k^e *represents the estimation error with distribution* $p_{x_k^e}$, *and* υ_k *denotes the measurement noise with distribution* p_{υ_k}. *Thus, the posterior particle set is expressed as:*

$$X_k^+ = [x_{k,r}^{1,-}, \cdots, x_{k,r}^{\gamma,-}, x_{k,\text{new}}^{1,-}, \cdots, x_{k,\text{new}}^{\eta,-}], \quad \gamma + \eta = N,$$

and the weights of the posterior particles are updated according to

$$w_k^i = w_{k-1}^i\, p(y_k \mid x_k^{i,+}).$$

Proof 5.4. According to Eqs. (5.100)–(5.106), at time k, the minimal-volume hyperparallel feasible domain of particles is defined as:

$$\mathcal{P}(\overline{x}_{k,k}, \overline{G}_{k,k}) = \left\{x : x = \overline{x}_{k,k} + \overline{G}_{k,k}\alpha, \ \|\alpha\|_\infty \le 1\right\}. \tag{5.109}$$

However, this formulation is not convenient for screening valid particles. By invoking Definition 5.4 and substituting $\overline{G}_{k,k} = P_k^{-1}$, an alternative representation of $\mathcal{P}(\overline{x}_{k,k}, \overline{G}_{k,k})$ is written as:

$$\begin{aligned}\mathcal{P}(\overline{x}_{k,k}, \overline{G}_{k,k}) &= \left\{x : x = \overline{x}_{k,k} + P_k^{-1}\alpha, \ \|\alpha\|_\infty \le 1\right\} \\ &= \left\{x : \|P_k(x - \overline{x}_{k,k})\|_\infty \le 1\right\}.\end{aligned} \tag{5.110}$$

When all prior particles X_k^-, sampled at time k according to the importance density function, are evaluated using Eq. (5.110), the validity condition is obtained as:

$$w_k^i = 0, \quad \text{if } P_k(x_k^{i,-} - \overline{x}_{k,k}) > 1. \tag{5.111}$$

Here, the retained particles are denoted as $x_{k,r}^{i,-}$, while the discarded ones are denoted as $x_{k,d}^{i,-}$. To maintain a constant total number of particles N, new effective particles $x_{k,\text{new}}^{i,-}$ must be generated. The maximum a posteriori (MAP) criterion is adopted to project particles lying outside the feasible domain back into the feasible region, thereby preserving their correlation with the underlying model. The state $\hat{x}_k$ is selected as the target parameter, and Y_k denotes the measurement dataset. From Eq. (5.8), the posterior probability density function of $\hat{x}_k$ is $p(x_k \mid Y_k)$. Based on the MAP criterion, the state estimation $\hat{x}_k$ is expressed as:

$$\begin{aligned}\hat{x}_k &= \arg\max_{x_k}\ p(x_k \mid Y_k) \propto \arg\max_{x_k}\ p(y_k \mid x_k)\, p(x_k \mid Y_{k-1}) \\ &= \arg\max_{x_k}\left\{p_{\upsilon_k}[y_k - h(x_k)]\right\} p_{x_k^e}(x_k - x_k^-),\end{aligned} \tag{5.112}$$

where p_{υ_k} and $p_{x_k^e}$ denote the distributions of measurement noise υ_k and estimation error x_k^e, respectively.

Accordingly, the particle feasible domain $\mathcal{P}(\overline{x}_{k,k}, \overline{G}_{k,k})$ serves as the constraint region for prior particles $x_k^{i,-}$. Substituting Eq. (5.112), the following constrained optimization problem is obtained:

$$\begin{aligned}\min_{x_{k,\text{new}}^{i,-} \in \mathcal{P}(\bar{x}_{k,k}, \bar{G}_{k,k})} \quad & -\Big(\lg[p_{x_k^e}(x_{k,\text{new}}^{i,-} - x_k^{i,-})] \\ & + \lg\{p_{\upsilon_k}[y_k - h(x_{k,\text{new}}^{i,-})]\}\Big), \\ \text{s.t.} \quad & x_{k,\text{new}}^{i,-} = x_k^{i,-} + x_k^e, \\ & y_k = h(x_{k,\text{new}}^{i,-}) + \upsilon_k, \\ & x_{k,\text{new}}^{i,-} \in \mathcal{P}(\bar{x}_{k,k}, \bar{G}_{k,k}),\end{aligned} \tag{5.113}$$

where x_k^e is the estimation error that follows the distribution $p_{x_k^e}$, and measurement noise υ_k follows the distribution p_{υ_k}. So posterior particles can be represented as:

$$X_k^+ = \{x_k^{i,+}\}_{i=1}^N = [x_{k,r}^{1,-}, \cdots, x_{k,r}^{\gamma,-}, x_{k,\text{new}}^{1,-}, \cdots, x_{k,\text{new}}^{\eta,-}], \quad \gamma + \eta = N,$$

and the weights of posterior particles are updated as, and the posterior particle weights are updated according to $w_k^i = w_{k-1}^i p(y_k | x_k^{i,+})$.

5.4.2.3 Summary of the P-IPF

The proposed improved particle filter algorithm based on the parallelotope (P-IPF) can be summarized as Figure 5.4.

5.4.3 Conclusion

In summary, the parallelotope-based improved particle filter (P-IPF) embeds the set-membership paradigm into the conventional particle filtering framework,

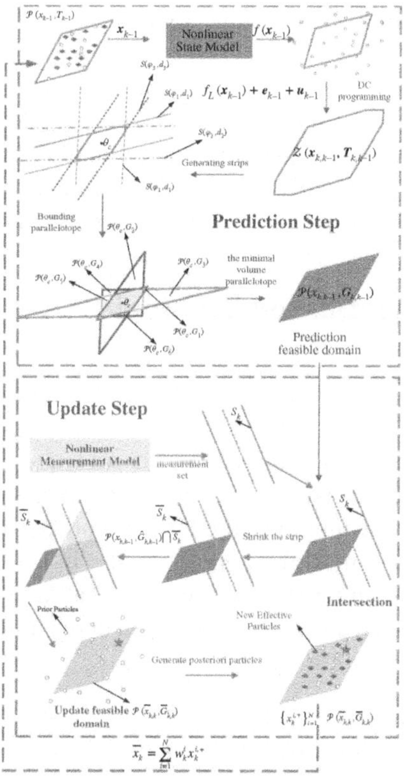

Figure 5.4 Flow chart of the P-IPF algorithm.

thereby strengthening both estimation accuracy and computational efficiency in nonlinear dynamic systems. The central idea is to generate, at each prediction and update stage, a minimal-volume parallelotope that tightly encloses the admissible state trajectories under bounded disturbances and measurement errors. This hyperparallel representation substitutes traditional probabilistic noise assumptions with deterministic set constraints, ensuring that the feasible domain always remains consistent with system uncertainties.

Within this bounded domain, the resampling and selection of particles are performed under an MAP strategy, which discards particles of low relevance while retaining and duplicating those that carry significant information. This mechanism preserves particle diversity, mitigates degeneracy, and guarantees that posterior samples remain confined within a physically meaningful feasible region. Moreover, the recursive construction of outward parallelotope bounds, combined with efficient feasibility verification, reduces the computational demand relative to classical set-membership particle filters and polytope-based approaches, without sacrificing robustness.

As a result, the P-IPF framework achieves a balanced trade-off between precision and scalability. By unifying hyperparallel set representations with particle-based inference, it provides a practical and reliable tool for state estimation in scenarios characterized by bounded uncertainties, and establishes a methodological basis that can be readily extended to high-dimensional or time-varying systems.

5.5 Projected Particle-confinement-based Zonotopic Space Filtering

5.5.1 Preliminaries

Several preliminary concepts are introduced as the foundation for subsequent analysis. An interval set $X = \{x \in \mathbb{R} : a \leq x \leq b\}$ is equivalently represented by $X = [a, b]$, which contains all real numbers x within the closed interval bounded by a and b. Among these interval sets, the unit interval is denoted by $\mathbf{B} = [-1, 1]$. Furthermore, let $\mathbf{B}^r$ denote an r-dimensional unit box, constructed by the Cartesian product of r unit intervals. A zonotope is an affine transformation of the unit hypercube, and its fundamental properties are summarized as follows.

Definition 5.7 Consider a vector $p \in \mathbb{R}^n$ and a matrix $H = \{h_1, h_2, \dots, h_r\} \in \mathbb{R}^{n \times r}$ $(r \geq n)$, a common representation of an r-dimensional zonotope $\mathcal{Z}$ is

$$\mathcal{Z} = p + \sum_{i=1}^{r} \beta_i h_i = p \oplus H\mathbf{B}^r \triangleq \langle p, H \rangle, \tag{5.114}$$

where $-1 \leq \beta_i \leq 1$, p is called the center, H is the generator matrix of $\mathcal{Z}$.

Property 6 Given a zonotope $\mathcal{Z} = p \oplus H\mathbf{B}^r \subset \mathbb{R}^n$ and a state transition matrix M, standard matrix multiplication yields

$$M \otimes \mathcal{Z} = (M \otimes p) \oplus (M \otimes H)\mathbf{B}^r. \tag{5.115}$$

Lemma 5.3 *[73] Consider a zonotope $\mathcal{Z} = p \oplus H\mathbf{B}^r \subset \mathbb{R}^n$ with an interval s $(n < s < r)$, the columns of H are reordered in non-increasing order of their Euclidean norms, and the resulting matrix is denoted as $\hat{H}$. The preceding $s - n$ columns are then denoted as $\hat{H}_T$. During iterative updates, we obtain*

$$\mathcal{Z} \subseteq p \oplus \left[\hat{H}_T \quad rs(Q)\right]\mathbf{B}^s \subset \mathbb{R}^n, \tag{5.116}$$

with $rs(Q)$ being a diagonal matrix defined as:

$$rs(Q)_{i,i} = \sum_{j=s-n+1}^{r} |\hat{H}_{i,j}|, \quad i = 1, \dots, n. \tag{5.117}$$

Lemma 5.4 *[74] Consider a zonotope $\mathcal{Z} = p \oplus H\mathbf{B}^r \subset \mathbb{R}^n$, together with a strip $S = \{x \in \mathbb{R}^n : |c^\top x - d| \leq \sigma\}$ and a vector $\lambda \in \mathbb{R}^n$, the smallest zonotopic boundary $\hat{\mathcal{Z}}$ that contains the intersection of $\mathcal{Z}$ and S can be determined as:*

$$\mathcal{Z} \cap S \subseteq \hat{\mathcal{Z}} = \hat{p} \oplus \hat{H}\mathbf{B}^{r+1}, \tag{5.118}$$

where

$$\hat{p} = p + \lambda(d - c^\top p), \qquad \hat{H} = \left[(I - \lambda c^\top)H \quad \sigma\lambda\right] \tag{5.119}$$

To minimize the F-norm of $\hat{H}$, the optimal λ is chosen as:

$$\lambda^* = \frac{\hat{H}\hat{H}^\top c}{c^\top \hat{H}\hat{H}^\top c + \sigma^2}. \tag{5.120}$$

Lemma 5.5 *[75] Given a zonotope $\mathcal{Z} = p \oplus H\mathbf{B}^r \subset \mathbb{R}^n$ that contains the real state x, the state is enclosed in the Box($\mathcal{X}$) interval $\mathcal{X} = [x^-, x^+]$, where the bounds x^+ and x^- can be calculated using the center p and generator matrix H as follows:*

$$\begin{cases} x_i^+ = p_i + \sum_{j=1}^{r} |H(i,j)|, & i = 1, \dots, n, \\ x_i^- = p_i - \sum_{j=1}^{r} |H(i,j)|, & i = 1, \dots, n, \end{cases} \tag{5.121}$$

where $H(i,j)$ denotes the element of H in the i-th row and the j-th column.

Lemma 5.6 *[76] Randomly initialize a particle swarm $\{P_{i,k}\}_{i=1}^{N}$, which consists of N particles in an m-dimensional search space. In the context of already searched individual optimal solution vector $p_{\text{best},i}^k \in \mathbb{R}^m$ and group optimal solution $g_{\text{best}}^k \in \mathbb{R}^m$, the subsequent speed $V_i^{k+1} \in \mathbb{R}^m$ and the position $X_i^{k+1} \in \mathbb{R}^m$ of the ith particle are computed through the rule seen in* (5.13) *and* (5.14).

5.5.2 Design of Projected Particle-confinement-basedZonotopic Space Filtering Algorithm

In this subsection, a zonotopic observer is constructed to derive a zonotope that contains the true state, which is established as the initial search space for the particle swarm. Particles are generated in this space and iteratively optimized, causing the particle swarm to scatter around the true state. The minimum zonotopic space enclosing all the particles after all iterations is then obtained, which leads to a more precise estimate of the state bounds.

5.5.2.1 Establishment of the Zonotopic Search Space

The zonotopic algorithm is incorporated into the set-membership filtering framework to approximate the system state vector at the initial step. Based on this, the feasible set derived from the zonotope, which encloses the system state vector, is utilized as the initial search domain for the particle swarm initialization.

According to Lemma 5.3, for the system in (5.4), the predicted vector $x_{k|k+1}$ of x_{k+1} from time k to $k+1$ is predicted as:

$$\begin{aligned} x_{k|k+1} &= Ax_k \oplus Bu_k \oplus w_k \\ &= A(p_k \oplus H_k B^r) \oplus Bu_k \oplus WB^{n_w} \\ &= Ap_k \oplus Bu_k \oplus AH_k B^r \oplus WB^{n_w} \\ &= Ap_k \oplus Bu_k \oplus [AH_k \quad W]B^{r+n_w}. \end{aligned} \tag{5.122}$$

We can then obtain the particle swarm prediction zonotopic space

$$Z_{k|k+1} = \langle p_{k|k+1}, H_{k|k+1} \rangle, \tag{5.123}$$

where

$$p_{k|k+1} = Ap_k \oplus Bu_k, \qquad H_{k|k+1} = [AH_k \quad W]. \tag{5.124}$$

Next, the predicted zonotopic space is updated according to the system output equation. This can be expressed in the form of a measurement strip

$$S_{k+1} = \{\theta \mid |Cx_{k+1} - y_{k+1}| \leq V\}. \tag{5.125}$$

According to Lemma 5.4, intersecting $Z_{k|k+1}$ with the measurement strip S_{k+1} yields a smaller region that contains x_{k+1} and determines its smallest zonotopic boundary $Z_{k+1} = \langle p_{k+1}, H_{k+1} \rangle$:

$$Z_{k|k+1} \cap S_{k+1} \subseteq Z_{k+1} = p_{k+1} \oplus H_{k+1}B^{r+1} \tag{5.126}$$

with

$$\begin{aligned} p_{k+1} &= p_{k|k+1} + \lambda^*(y_{k+1} - Cx_{k+1}) \\ &= Ap_k \oplus Bu_k + \lambda^*(y_{k+1} - Cx_{k+1}), \end{aligned} \tag{5.127}$$

$$\begin{aligned} H_{k+1} &= \left[(I - \lambda^* C)H_{k|k+1} \quad V\lambda^*\right] \\ &= \left[(I - \lambda^* C)AH_k \quad (I - \lambda^* C)W \quad V\lambda^*\right], \end{aligned} \tag{5.128}$$

$$\lambda^* = H_{k|k+1}H_{k|k+1}^T C^T \left(CH_{k|k+1}H_{k|k+1}^T C^T + V^2\right)^{-1}. \tag{5.129}$$

From this derivation, we obtain the zonotopic space Z_{k+1} that contains x_{k+1}, which is then established as the search space for subsequent initialization and iterative optimization of the particle swarm.

5.5.2.2 Projected Particle-confinement-based Zonotopic Space Update

Upon acquiring the zonotopic space Z_{k+1}, the particle swarm engages in an iterative optimization search centered around the actual state x_{k+1} within it at the $k+1$ time step before initializing and starting optimization search of particle swarm $\{P_{i,k+1}\}_{i=1}^{N}$ at each iteration. To ensure that the particle does not exceed the search space after initialization and before each iteration optimization, a feasible method must be studied.

Theorem 5.5 *For the known zonotopic search space $Z_{k+1} = \langle p_{k+1}, H_{k+1}\rangle$ at $k+1$, at the beginning of iteration q of the particle swarm $\{P_{i,k+1}\}_{i=1}^{N}$, the judgment condition that the i-th particle's coordinate $P_{i,k+1}^{q}$ in the particle swarm exceeds the search space in the search process is*

$$||d_{k+1}|| \leq \left(\frac{d_{k+1}}{||d_{k+1}||}\right)^T p_{k+1} + \sum_{j=1}^{r}\left|\left(\frac{d_{k+1}}{||d_{k+1}||}\right)^T h_j\right|, \tag{5.130}$$

where

$$d_{k+1} = P_{i,k+1}^{q} - p_{k+1}, \tag{5.131}$$

and

$$H_{k+1} = [h_1, h_2, \dots, h_r]. \tag{5.132}$$

Proof 5.5. According to the related concepts of support functions in convex optimization problems, if a direction vector u is given, the expression for the support function for zonotope $Z_{k+1} = \{x \mid x = p_{k+1} + H_{k+1}\mathbf{B}^r,\ \mathbf{B} = [-1, 1]\}$ can be expressed as:

$$h_Z(u) = \sup_{x\in Z_{k+1}} (u \cdot x) = \sup_{\beta\in\mathbf{B}} \langle u \cdot p_{k+1} + H_{k+1}\beta^r\rangle, \tag{5.133}$$

where $\langle u \cdot p_{k+1} + H_{k+1}\beta^r\rangle$ represents the inner product of the direction u and vector $p_{k+1} + H_{k+1}\beta^r$.

For any direction vector u, the support function $h_Z(u)$ calculates the maximum value of the zone in that direction. The problem can thus be reformulated into the following optimization form:

$$\begin{aligned} &\max\ \langle u \cdot p_{k+1} + H_{k+1}\beta^r\rangle \\ &\text{s.t.}\quad \beta \in [-1, 1]. \end{aligned} \tag{5.134}$$

To simplify this optimization problem, the inner product can be expanded as:

$$\begin{aligned}\langle u \cdot p_{k+1} + H_{k+1}\beta^T \rangle &= \langle u \cdot p_{k+1} \rangle + \langle u \cdot H_{k+1}\beta^T \rangle \\ &= \langle u \cdot p_{k+1} \rangle + \left\langle u \cdot \sum_{j=1}^{r} \beta_j h_j \right\rangle \\ &= \langle u \cdot p_{k+1} \rangle + \sum_{j=1}^{r} \beta_j \langle u \cdot h_j \rangle,\end{aligned} \tag{5.135}$$

where $\langle u, p_{k+1} \rangle$ is a constant that does not affect the optimization. Therefore, the convex optimization problem in 5.134 can be simplified as:

$$\begin{aligned}\max \quad & \sum_{j=1}^{r} \beta_j \langle u \cdot h_j \rangle \\ \text{s.t.} \quad & \beta_j \in [-1, 1].\end{aligned} \tag{5.136}$$

Because each β_j lies within the interval $[-1, 1]$, the optimization can be separated across dimensions. Thus, the solution of 5.136 is

$$\sup_{\beta_j \in B} \beta_j \langle u \cdot h_j \rangle = \sum_{j=1}^{r} \left| \langle u \cdot h_j \rangle \right|. \tag{5.137}$$

We can therefore rewrite the support function $h_{\mathcal{Z}}(u)$ as:

$$\begin{aligned}h_{\mathcal{Z}}(u) &= \langle u \cdot p_{k+1} \rangle + \sum_{j=1}^{r} \left| \langle u \cdot h_j \rangle \right| \\ &= u^T p_{k+1} + \sum_{i=1}^{r} \left| u^T h_i \right|,\end{aligned} \tag{5.138}$$

which specifies the distance from the center p_{k+1} to the farthest boundary of $\mathcal{Z}_{k+1}$ along the direction u.

For particle $P^q_{i,k+1}$, we select the direction as:

$$u = \frac{d_{k+1}}{||d_{k+1}||} = \frac{P^q_{i,k+1} - p_{k+1}}{||P^q_{i,k+1} - p_{k+1}||}. \tag{5.139}$$

Therefore, the condition for $P^q_{i,k+1}$ to remain inside $\mathcal{Z}_{k+1}$ is equivalent to requiring the length of vector d_{k+1} not to exceed the value of the support function $h_{\mathcal{Z}}(d_{k+1})$, expressed as:

$$\begin{aligned}||P^q_{i,k+1} - p_{k+1}|| = ||d_{k+1}|| &\le h_{\mathcal{Z}}\left(\frac{d_{k+1}}{||d_{k+1}||} \right) \\ &= \left(\frac{d_{k+1}}{||d_{k+1}||} \right)^T p_{k+1} + \sum_{i=1}^{r} \left| \left(\frac{d_{k+1}}{||d_{k+1}||} \right)^T h_j \right|.\end{aligned} \tag{5.140}$$

Thus, we can utilize Theorem 1 to ascertain whether the particle surpasses the search space following initialization and throughout each iteration of optimization. Furthermore, when the i-th particle's coordinate $P^q_{i,k+1}$ exceeds the confines of the search space, as:

$$P^q_{i,k+1} \notin \mathcal{Z}_{k+1} = \langle p_{k+1}, H_{k+1} \rangle, \tag{5.141}$$

we may proceed to update and modify its coordinates as follows.

Theorem 5.6 *Given d_{k+1} in 5.131, the optimal updated position of $P^q_{i,k+1}$ that satisfies the property of not exceeding the search boundary $\mathcal{Z}_{k+1}$ while preserving its original position as much as possible is given by*

$$P^q_{i,k+1}{}' = p_{k+1} + H_{k+1}\beta^r_*. \tag{5.142}$$

where β^r_* is derived by addressing the quadratic programming problem:

$$\begin{aligned} \min \quad & d_k^T d_k - 2d_k^T H_{k+1}\beta^r + \beta^{rT} H_{k+1}^T H_{k+1}\beta^r \\ s.t. \quad & \beta \in [-1, 1]. \end{aligned} \tag{5.143}$$

Proof 5.6. First, we can construct an optimization problem to minimize the distance between $P^q_{i,k+1}$ and points in $\mathcal{Z}_{k+1}$:

$$\min_{\beta\in[-1,1]} \quad \|p_{k+1} + H_{k+1}\beta - P^q_{i,k+1}\|^2. \tag{5.144}$$

Because $d_{k+1} = P^q_{i,k+1} - p_{k+1}$, the objective function becomes

$$\begin{aligned} \|d_{k+1} + H_{k+1}\beta^r\|^2 &= (d_{k+1} + H_{k+1}\beta^r)^T (d_{k+1} + H_{k+1}\beta^r) \\ &= d_k^T d_k - 2d_k^T H_{k+1}\beta^r + \beta^{rT} H_{k+1}^T H_k \beta^r. \end{aligned} \tag{5.145}$$

In summary, the entire optimization problem can be reformulated as:

$$\begin{aligned} \min \quad & d_k^T d_k - 2d_k^T H_{k+1}\beta^r + \beta^{rT} H_{k+1}^T H_{k+1}\beta^r \\ \text{s.t.} \quad & \beta \in [-1, 1]. \end{aligned} \tag{5.146}$$

The optimal solution β^* for β is acquired through solving the above optimization problem. Then, by substituting the optimal solution into the definition of $\mathcal{Z}_{k+1}$, the optimal projection position $P^q_{i,k+1}{}'$ in $\mathcal{Z}_{k+1}$ can be obtained as follows:

$$P^q_{i,k+1}{}' = p_{k+1} + H_{k+1}\beta^r_*. \tag{5.147}$$

Constraint $\beta \in [-1, 1]^r$ ensures that the projected point $P^q_{i,k+1}{}'$ falls within the defined range of $\mathcal{Z}_{k+1}$, thereby ensuring the validity and correctness of the projection results, thus preserving the characteristics of its original position as much as possible and avoiding excessive correction of particle positions during the projection process.

We take the real state as the optimal solution that the PSO algorithm needs to find and take the position of each particle as the potential solution of the state. During iteration q, each potential solution $P^q_{i,k+1}$ of the particle swarm $\{P^q_{i,k+1}\}_{i=1}^{N}$ can serve as a possible estimate of the current system state, and by substituting it into the system equation, the corresponding system observation estimate $\hat{y}^q_{i,k+1}$ for each potential solution can be obtained as:

$$\hat{y}^q_{i,k+1} = CP^q_{i,k+1} + \hat{v}_k, \tag{5.148}$$

where $\hat{v}_k$ is a random quantity in $\langle 0, V \rangle$.

Because $\hat{y}^q_{i,k+1}$ and the real system output y_{k+1} are known quantities containing $P^q_{i,k+1}$ and x_{k+1}, respectively, a fitness function is defined as follows to evaluate the quality of unknown particles:

$$\text{fitness}(P^q_{i,k+1}) = \exp\left[-\left\|y_{k+1} - \hat{y}^q_{i,k+1}\right\|_2^2\right]. \tag{5.149}$$

The position where the i-th particle has the highest corresponding fitness function that has occurred is recorded as $P^k_{\text{best},i}$, and the position with the highest corresponding fitness function that has occurred among all particles is denoted as g^k_{best}. During the q-th to $(q+1)$-th iterations of the particle swarm, according to Lemma 5.6, the velocity and position of each particle are updated as follows:

$$V^{q+1}_{i,k+1} = \omega V^q_{i,k+1} + c_1 r_1\left(P^k_{\text{best},i} - P^q_{i,k+1}\right) + c_2 r_2\left(g^k_{\text{best}} - P^q_{i,k+1}\right), \tag{5.150}$$

$$P^{q+1}_{i,k+1} = P^q_{i,k+1} + V^{q+1}_{i,k+1}. \tag{5.151}$$

According to Theorems 5.5 and 5.6, we can ensure that particles do not exceed the search space boundary before each iteration. After each state update, particles must reevaluate their fitness values and compare them with the best-known values of $P^k_{\text{best},i}$ and g^k_{best}. Based on actual requirements, an iteration termination threshold δ can be set for the particle swarm. When the global optimal value reaches or surpasses the preset iteration termination threshold, the particle swarm is distributed near the actual state. The particle swarm at this instant is recorded as:

$$\{P_{i,k+1}\}_{i=1}^{N} = \{P_{1,k+1}, P_{2,k+1}, \cdots, P_{N,k+1}\}. \tag{5.152}$$

In order to enhance the precision of state estimation, we aim to more tightly encapsulate the true state x_{k+1} and the entire particle swarm, thereby narrowing $\mathcal{Z}_{k+1}$. We denote this narrowed $\mathcal{Z}_{k+1}$ as $\bar{\mathcal{Z}}_{k+1}$, representing the more precise feasible set for the zonotopic algorithm at time $k+1$. In this study, we further explore a method for determining the center and generative matrix of $\bar{\mathcal{Z}}_{k+1}$.

Theorem 5.7 *Given that the randomly initialized particle swarm* $\{P^0_{i,k+1}\}_{i=1}^N$ *generated at* $k+1$ *is scattered within a zonotopic space* $\mathcal{Z}_{k+1} = \langle p_{k+1}, H_{k+1}\rangle$ *containing the true state, after iterative optimization, the corresponding estimation bounds of* x_{k+1} *are calculated as follows:*

$$\begin{cases} x^+_{k+1}(i) = p_{k+1}(i) + \sum_{j=1}^{r} |\bar{H}_{k+1}(i,j)|, & i = 1, \dots, N, \\ x^-_{k+1}(i) = p_{k+1}(i) - \sum_{j=1}^{r} |\bar{H}_{k+1}(i,j)|, & i = 1, \dots, N, \end{cases} \tag{5.153}$$

where

$$\bar{H}_{k+1} = \{\bar{h}_1, \bar{h}_2, \dots, \bar{h}_r\}, \tag{5.154}$$

$$\bar{h}_j = \frac{h_j}{1 + \text{step}_j \parallel h_j \parallel}, \quad j = 1, 2, \dots, r, \tag{5.155}$$

$$\text{step}_j = \min\{d^{k+1}_{\text{pos}}, d^{k+1}_{\text{neg}}\}, \tag{5.156}$$

$$d^{k+1}_{\text{pos}} = \min\left\{\frac{\left|h_j^T(P_{i,k+1} - p_{k+1}) - 1\right|}{\parallel h_j \parallel}\right\}, \quad i = 1, 2, \dots, N, \tag{5.157}$$

$$d^{k+1}_{\text{neg}} = \min\left\{\frac{\left|h_j^T(P_{i,k+1} - p_{k+1}) + 1\right|}{\parallel h_j \parallel}\right\}, \quad i = 1, 2, \dots, N. \tag{5.158}$$

Proof 5.7. For a zonotopic space $\mathcal{Z}_{k+1}$ defined by the center point p_{k+1} and the generator matrix $H_{k+1} = \{h_1, h_2, \dots, h_r\}$, its boundary is composed of the following $2r$ hyperplanes:

$$\{h_j^T x = h_j^T p_{k+1} + 1, \ h_j^T x = h_j^T p_{k+1} - 1\}_{j=1}^r, \tag{5.159}$$

where each pair of hyperplanes is symmetric about p_{k+1} and can be divided into a positive hyperplane $h_j^T x = h_j^T p_{k+1} + 1$ and a negative hyperplane $h_j^T x = h_j^T p_{k+1} - 1$. The general vector h_j serves as their normal vector.

For the particle swarm $\{P_{i,k+1}\}_{i=1}^N$ and a pair of hyperplanes, the distances from each particle $P_{i,k+1}$ to the positive and negative hyperplanes are given as:

$$d_{\text{pos}} = \frac{|h_j^T(P_{i,k+1} - p_{k+1}) - 1|}{\|h_j\|}, \tag{5.160}$$

$$d_{\text{neg}} = \frac{|h_j^T(P_{i,k+1} - p_{k+1}) + 1|}{\|h_j\|}. \tag{5.161}$$

Therefore, for the same set of hyperplanes, the particle closest to the positive hyperplane and the particle closest to the negative hyperplane can be determined. The minimum distances corresponding to these two particles are:

$$d_{\text{pos}}^{k+1} = \min\left\{\frac{|h_j^T(P_{i,k+1} - p_{k+1}) - 1|}{\|h_j\|}\right\}, \qquad i = 1, 2, \dots, N, \tag{5.162}$$

$$d_{\text{neg}}^{k+1} = \min\left\{\frac{|h_j^T(P_{i,k+1} - p_{k+1}) + 1|}{\|h_j\|}\right\}, \qquad i = 1, 2, \dots, N. \tag{5.163}$$

Once these distances are determined, one can move the set of hyperplanes along the normal vector h_j toward the center point p_{k+1}, and select the moving step size as:

$$\text{step}_j = \min\{d_{\text{pos}}^{k+1}, \qquad d_{\text{neg}}^{k+1}\}, \tag{5.164}$$

which ensures that the overall particle swarm remains confined between the positive and negative hyperplanes. If one scales the normal vectors of r pairs of hyperplanes enclosed in $\mathcal{Z}_{k+1}$ in this manner, the scaled normal vector is represented as:

$$\bar{h}_j = \frac{h_j}{1 + step_j\|h_j\|}, \qquad j = 1, 2, \dots, r. \tag{5.165}$$

The shifted r pairs of hyperplanes form a zonotopic space $\bar{\mathcal{Z}}_{k+1}$ that more tightly encloses $\{P_{i,k+1}\}_{i=1}^{N}$ and the real state x_{k+1}. Because each pair of positive and negative hyperplanes moves in equal steps, the center of $\bar{\mathcal{Z}}_{k+1}$ is consistent with p_{k+1}, and the generation matrix is composed of the scaled generator vectors $\bar{h}_j$, denoted as:

$$\{P_{i,k+1}\}_{i=1}^{N} \in \bar{\mathcal{Z}}_{k+1} = \langle \bar{p}_{k+1}, \bar{H}_{k+1} \rangle, \tag{5.166}$$

where

$$\bar{p}_{k+1} = p_{k+1}, \tag{5.167}$$

$$\bar{H}_{k+1} = \{\bar{h}_1, \bar{h}_2, \dots, \bar{h}_r\}. \tag{5.168}$$

$\bar{\mathcal{Z}}_{k+1}$ is a more compact zonotopic space containing the particle swarm $\{P_{i,k+1}\}_{i=1}^{N}$. According to Lemma 5.5, the bounds of the state estimation x_{k+1} are determined as follows:

$$\begin{cases} x_{k+1}^{+}(i) = p_{k+1}(i) + \sum_{j=1}^{r} |\bar{H}_{k+1}(i, j)|, & i = 1, \dots, N, \\ x_{k+1}^{-}(i) = p_{k+1}(i) - \sum_{j=1}^{r} |\bar{H}_{k+1}(i, j)|, & i = 1, \dots, N, \end{cases} \tag{5.169}$$

where $\bar{H}_{k+1}(i,j)$ corresponds to the element of $\bar{H}_{k+1}$ located at the i-th row and j-th column.

By taking the two-dimensional space as an illustrative example, Figure 5.5 demonstrates the process of optimizing and updating the zonotopic search space at $k+1$ using the PPC-ZSF algorithm. Initially, particles are uniformly distributed inside the zonotopic search space $\mathcal{Z}_{k+1} = \langle p_{k+1}, H_{k+1} \rangle$ and subjected to boundary checks, where boundary projection is applied if any particles exceed the feasible region to ensure all particles remain confined within it. Guided by the fitness function, particles subsequently adjust their velocities and positions through both individual and group experiences during each iteration, thereby gradually converging toward the global optimal solution. After multiple iterations, the particle swarm approaches the true optimal state, marking the completion of the optimization

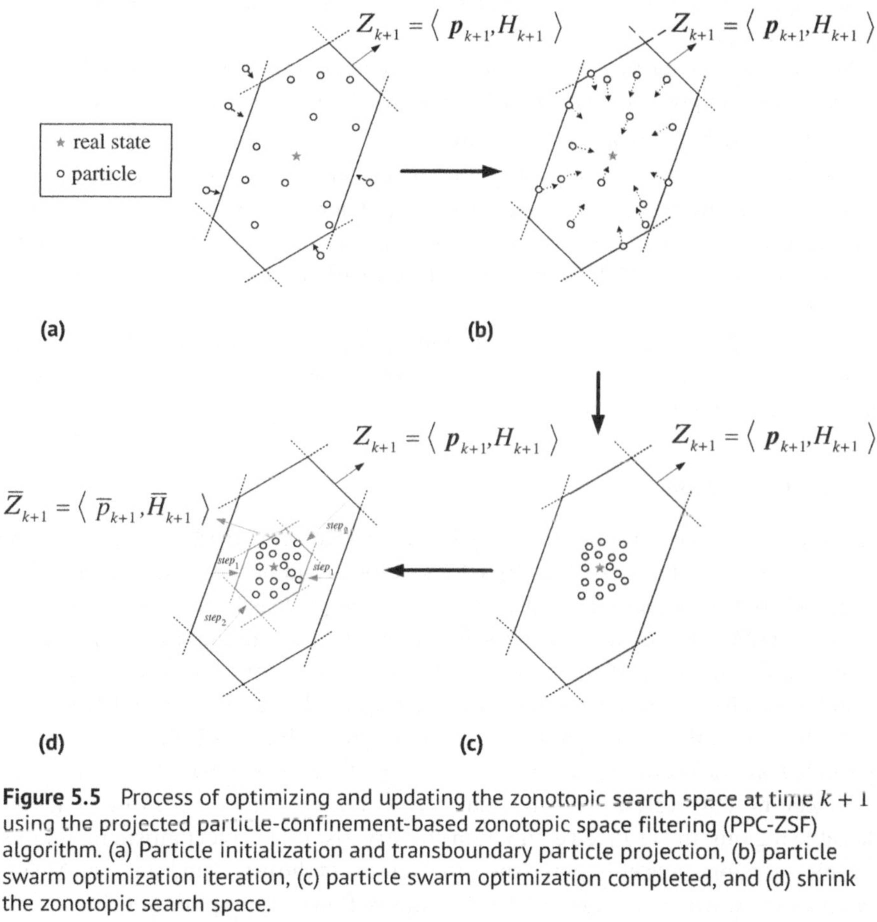

Figure 5.5 Process of optimizing and updating the zonotopic search space at time $k+1$ using the projected particle-confinement-based zonotopic space filtering (PPC-ZSF) algorithm. (a) Particle initialization and transboundary particle projection, (b) particle swarm optimization iteration, (c) particle swarm optimization completed, and (d) shrink the zonotopic search space.

process. Finally, by scaling the normal vector to adjust the hyperplane, the optimized and updated zonotopic search space $\bar{\mathcal{Z}}_{k+1} = \langle p_{k+1}, \bar{H}_{k+1} \rangle$ is generated, which significantly improves the precision of state estimation.

In summary, the PPC-ZSF algorithm could be outlined through the following steps:

Step 1: Initialize $k = 0$ with the initial search space $\mathcal{Z}_0 = \langle p_0, H_0 \rangle$, and set the total sample time K.

Step 2: Use (5.124) to compute the predicted state space $\mathcal{Z}_{k|k+1} = \langle p_{k|k+1}, H_{k|k+1} \rangle$, then compute the intersection with the measurement set S_{k+1} using (5.126) to obtain the updated search space $\mathcal{Z}_{k+1}$, which serves as the basis for particle swarm initialization and optimization.

Step 3: Initialize a particle swarm with n particles in $\mathcal{Z}_{k+1}$. Check the boundary conditions using (5.130) before each swarm update. If any particles exceed the search space, perform boundary projections by solving the quadratic programming problem in (5.142).

Step 4: For each particle within the search space, compute its fitness and update the velocity and position using (5.150) and (5.151). This yields the particle swarm $\{P^q_{i,k+1}\}_{i=1}^N$ distributed near the true state.

Step 5: Move each pair of hyperplanes as shown in (5.159) to form the search space, with the movement step size of each pair determined from (5.162) to (5.164). Record the scaling factor of the normal vectors of these hyperplanes, as given in (5.165), and then use (5.167) and (5.168) to calculate the compressed zonotopic search space $\bar{\mathcal{Z}}_{k+1}$.

Step 6: Repeat Step 2 for each iteration until $k = K$, and output the extremal bounds together with the center of the zonotopic space, which serve as the system state estimation result.

5.5.3 Conclusion

In summary, the proposed PPC-ZSF algorithm effectively integrates set-membership filtering with PSO to achieve robust and accurate state estimation under UBB disturbances. By constructing a zonotopic feasible region as the particle search space, the method ensures that all particles are initialized and iteratively updated within bounded geometric constraints. A boundary detection and projection mechanism is employed to identify and optimally re-map out-of-bound particles back into the zonotopic region, thereby preserving feasibility and avoiding particle loss. Furthermore, a hyperplane contraction strategy based on scaled normal vectors adaptively adjusts the boundaries of the zonotope, resulting in tighter feasible regions and reduced estimation conservatism. This mechanism maintains geometric rigor while enhancing the compactness of the estimated state set. Meanwhile, the swarm dynamics guided by fitness functions exploit both local and global information, accelerating convergence toward accurate state estimates.

Overall, the PPC-ZSF algorithm achieves a balanced trade-off between robustness and precision. It guarantees that particles remain confined within the zonotopic feasible region throughout the optimization process, reduces redundant search volume, and improves the efficiency of state estimation. These features highlight its potential applicability to high-dimensional and dynamically varying systems where reliable state estimation is critical.

5.6 Zonotopic Feasible Set Optimized Filter Based on Differential Evolution

5.6.1 Preliminaries

This subsection introduces notations and fundamental properties. The real numbers set is represented by $\mathbb{R}$, while $\mathbb{R}^d$ indicates the d-dimensional Euclidean space.

Definition 5.8 [77] A *zonotope* $\mathcal{Z} \subset \mathbb{R}^d$ with order q and dimension d ($d \leq q$) is described as:

$$\mathcal{Z} = p \oplus GB^q = \{z \in \mathbb{R}^d \mid z = p + G\beta,\ \beta \in B^q\}, \tag{5.170}$$

where $B^q = [-1, 1]^q$ defines a q-dimensional box, $p \in \mathbb{R}^d$ is the center, $G \in \mathbb{R}^{d\times q}$ is the generator matrix. The interval hull of $\mathcal{Z} = \langle p, G\rangle$ is

$$\mathcal{Z}_I = [z^-, z^+] = \{z \in \mathbb{R}^d : z^- \leq z \leq z^+\}. \tag{5.171}$$

Definition 5.9 [77] The Minkowski sum of $\mathcal{Z}_1 = \langle p_1, G_1\rangle$ and $\mathcal{Z}_2 = \langle p_2, G_2\rangle$ is given by

$$\mathcal{Z}_1 \oplus \mathcal{Z}_2 = \langle p_1 + p_2,\ [G_1\ \ G_2]\rangle, \tag{5.172}$$

and a linear transformation $L \in \mathbb{R}^{d\times q}$ applied to $\mathcal{Z}$ yields

$$L \otimes \langle p, G\rangle = \langle Lp,\ LG\rangle. \tag{5.173}$$

Definition 5.10 [78] The i-th component of the interval is

$$z_i^- = p_i - \sum_{j=1}^{q} |G_{i,j}|, \qquad z_i^+ = p_i + \sum_{j=1}^{q} |G_{i,j}|, \tag{5.174}$$

where z_i^-, z_i^+, and p_i denote the i-th entries of z^-, z^+, and p, separately.

Definition 5.11 [78] A zonotope $\mathcal{Z} = \langle p, G\rangle \subset \mathbb{R}^d$ ($d \leq s \leq q$) may be represented through an approximation using a lower-order zonotope:

$$\mathcal{Z} \approx \langle p, G_{\downarrow s}\rangle, \tag{5.175}$$

where columns of G are sorted by norm in descending order, and split into $G_{\geq}$ (largest $s-d$ columns) and $G_{<}$ (remaining columns). $G_{<}$ is replaced with its orthometric projection $qs(G_{<})$, to yield:

$$G_{\downarrow s} = [G_{\geq},\ qs(G_{<})] \in \mathbb{R}^{d\times s}. \tag{5.176}$$

Definition 5.12 [78] Some useful quantities of the zonotopic space analysis include:

- The Frobenius radius of $\mathcal{Z} = \langle p, G\rangle$ is

$$||\langle p, G\rangle||_F = ||G||_F = \sqrt{\mathrm{tr}(G^T G)} = \sqrt{\mathrm{tr}(GG^T)}. \tag{5.177}$$

When considering a weighting matrix M, it becomes

$$||G||_F^M = \sqrt{\mathrm{tr}(G^T MG)}. \tag{5.178}$$

- The covariance of $\mathcal{Z}$ is

$$\mathrm{cov}(p, G) = GG^T. \tag{5.179}$$

Property 7 Trace operator properties [79]:

$$\begin{aligned}
\mathrm{tr}(A) &= \mathrm{tr}(A^{\mathrm{T}}),\\
\mathrm{tr}(AB) &= \mathrm{tr}(BA),\\
\frac{\partial \mathrm{tr}(AX^{\mathrm{T}}B)}{\partial X} &= A^{\mathrm{T}}B^{\mathrm{T}},\\
\frac{\partial \mathrm{tr}(AXBX^{\mathrm{T}}C)}{\partial X} &= B^{\mathrm{T}}X^{\mathrm{T}}CA + B^{\mathrm{T}}X^{\mathrm{T}}A^{\mathrm{T}}C^{\mathrm{T}}.
\end{aligned} \tag{5.180}$$

5.6.1.1 Differential Evolution Algorithm

Differential evolution (DE) is an evolutionary algorithm and operates on a population that iteratively refines candidate solutions through mutation, crossover, and selection [80].

a) Mutation: In each generation, a mutant vector is generated by combining three distinct individuals from the population. Specifically, the difference of two vectors is scaled and added to the third:

$$v_i = x_{r_1} + \alpha \cdot (x_{r_2} - x_{r_3}), \tag{5.181}$$

where $x_{r_1}, x_{r_2}, x_{r_3}$ are randomly chosen distinct individuals, and $\alpha \in [0.5, 1]$ is the mutation factor.

b) Crossover: A trial vector u_i is constructed by mixing the mutant vector v_i with the current individual x_i. Each element of u_i is selected from v_i with probability CR, and from x_i otherwise:

$$u_{i,j} = \begin{cases} v_{i,j}, & \text{if } \mathrm{rand}_j < CR, \\ x_{i,j}, & \text{otherwise,} \end{cases} \tag{5.182}$$

where rand_j denotes a random variable uniformly drawn from interval $[0, 1]$.

c) Selection: The trial vector u_i is evaluated against the current individual x_i. If u_i yields better fitness than x_i, then u_i replaces x_i; otherwise, x_i is retained.

5.6.2 Main Conclusion

5.6.2.1 Design of the Zonotopic Kalman Filter

To achieve accurate state estimation of system (5.4) under UBB noise disturbances, a zonotopic Kalman filter (ZKF) is designed as:

$$\hat{x}(k+1) = A\hat{x}(k) + Bu(k) + Dw(k) + L(k)\left[y(k) - H\hat{x}(k) - Eu(k) - Fv(k)\right], \tag{5.183}$$

where $L(k)$ denotes the optimal zonotopic Kalman gain matrix, and $\hat{x}(k)$ is the estimated state at k instant.

Assume that at instant k, the predicted state is contained within $\hat{\mathbf{X}}(k) = \langle \hat{p}(k), \hat{G}(k) \rangle$. Then, from (5.183), we obtain:

$$\begin{aligned} \hat{x}(k+1) &= (A - L(k)H)\hat{x}(k) + Bu(k) + Dw(k) \\ &\quad - L(k)Eu(k) - L(k)Fv(k) \\ &\in (A - L(k)H)\langle \hat{p}(k), \hat{G}(k) \rangle \oplus Bu(k) \oplus D\mathcal{W}(k) \\ &\oplus (-L(k)F)\mathcal{V}(k) \\ &\subseteq \left\langle \begin{array}{c} (A - L(k)H)\hat{p}(k) + Bu(k) + L(k)y(k) - L(k)Eu(k), \\ (A - L(k)H) \downarrow \hat{G}(k) \oplus DG_w \oplus (-L(k)F) \downarrow G_v \end{array} \right\rangle, \end{aligned} \tag{5.184}$$

where the updated center and generator matrices are given by:

$$\begin{aligned} \hat{p}(k+1) &= (A - L(k)H)\hat{p}(k) + Bu(k) + L(k)\big(y(k) - Eu(k)\big), \\ \hat{G}(k+1) &= \left[(A - L(k)H) \downarrow \hat{G}(k) \quad DG_w \quad -L(k)FG_v\right]. \end{aligned}$$

The volume of the updated zonotope $\hat{\mathbf{X}}(k+1)$ can be quantified through the trace of its shape matrix. Accordingly, the objective function for estimation error covariance is formulated as:

$$J_M(k+1) = \mathrm{tr}\left(\hat{G}^{\mathrm{T}}(k+1) M \hat{G}(k+1)\right), \tag{5.185}$$

where M is a positive semi-definite weighting matrix. To determine the optimal gain $L(k)$, the partial derivative of $J_M(k+1)$ with respect to $L(k)$ is set to zero:

$$\frac{\partial J_M(k+1)}{\partial L(k)} = 0. \tag{5.186}$$

After simplification, the derivative is written as:

$$\frac{\partial J_M(k+1)}{\partial L(k)} = \frac{\partial}{\partial L(k)} \operatorname{tr}\left(ML(k)H\tilde{P}(k)H^{\mathrm{T}}L(k)^{\mathrm{T}} + ML(k)Q_{\upsilon}L(k)^{\mathrm{T}}\right). \tag{5.187}$$

Setting the derivative to zero yields:

$$\begin{aligned} &\frac{\partial}{\partial L(k)} \operatorname{tr}\left(M\left(L(k)H\tilde{P}(k)H^{\mathrm{T}}L(k)^{\mathrm{T}} + L(k)Q_{\upsilon}L(k)^{\mathrm{T}}\right)\right) \\ &= 2\frac{\partial}{\partial L(k)} \operatorname{tr}\left(MA\tilde{P}(k)H^{\mathrm{T}}L(k)^{\mathrm{T}}\right). \end{aligned} \tag{5.188}$$

Thus, the optimal gain matrix is obtained as:

$$L(k)S(k) = AR(k), \tag{5.189}$$

where $S(k) = H\tilde{P}(k)H^{\mathrm{T}} + Q_{\upsilon}$, $R(k) = \tilde{P}(k)H^{\mathrm{T}}$.

5.6.2.2 DE-ZKF-based Generator Matrix Optimization

As discussed in Section 5.6.2.1, the conventional ZKF filter often exhibits conservativeness, producing zonotopes with overly wide boundaries. This leads to reduced estimation accuracy and weaker responsiveness to disturbances and noise.

To mitigate this issue, a differential evolution (DE) algorithm is integrated to optimize the generator matrix $\hat{G}(k)$, thereby achieving effective boundary compression and reducing redundancy. The population at each time step k could be represented by:

$$\mathrm{Pop}(k) = \{G_i(k) \mid i = 1, 2, \ldots, N\}, \tag{5.190}$$

where N denotes the population size, and each individual $G_i(k)$ corresponds to a candidate generator matrix.

Each individual is initialized by perturbing the current generator matrix:

$$G_i(k) = \hat{G}(k) + \delta_i, \tag{5.191}$$

where δ_i is a perturbation drawn from a standard normal distribution:

$$\delta_i = \mathrm{randn}(\mathrm{size}(\hat{G}(k))) \times \sigma, \tag{5.192}$$

with σ being the perturbation scaling factor. The operator $\mathrm{randn}(\mathrm{size}(\hat{G}(k)))$ generates a standard normal random matrix of the same dimensions as $\hat{G}(k)$.

During each generation of DE, the fitness of every generator matrix $G_{i,t}(k)$ is evaluated. The fitness function jointly considers the zonotope volume, coverage error, and a penalty term to ensure compactness and accuracy. The zonotope volume is approximated by the Frobenius norm:

$$\mathrm{Vol}(G_{i,t}(k)) = ||\hat{G}_{i,t}(k)||_F, \tag{5.193}$$

where the Frobenius norm $|| \cdot ||_F$ is employed to measure the overall magnitude of the generator matrix. The coverage error characterizes the deviation between the estimated and true outputs, thereby reflecting estimation accuracy. Under the ZKF framework, the estimated value $\hat{y}(k)$ associated with $\hat{x}(k)$ satisfies:

$$\begin{aligned}
\hat{y}(k) &= H\hat{x}(k) + Bu(k) + Fv(k) \\
&= (H\langle \hat{p}(k), \hat{G}(k)\rangle) \oplus Bu(k) \oplus (F \odot \langle 0, G_v\rangle) \\
&\subseteq \langle H\hat{p}(k) + Bu(k), [H\hat{G}(k) \;\; FG_v]\rangle \\
&= \langle \hat{p}_y(k), \hat{G}_y(k)\rangle,
\end{aligned} \tag{5.194}$$

where the center is $\hat{p}_y(k) = H\hat{p}(k) + Bu(k)$, and the generator $\hat{G}_y(k) = [H\hat{G}(k) \;\; FG_v]$ accounts for output uncertainty.

For the i-th individual, the coverage error is formulated as:

$$\begin{aligned}
e(\hat{G}_{i,t}(k)) = max\Bigg\{ &\left\| y(k) - \left(\hat{p}_y(k) + \sum_{l=1}^{q} \hat{G}_{y,i}(:, l) \right) \right\|_1, \\
&\left\| y(k) - \left(\hat{p}_y(k) - \sum_{l=1}^{q} \hat{G}_{y,i}(:, l) \right) \right\|_1 \Bigg\}, \\
&i = 1, 2, \dots, N,
\end{aligned} \tag{5.195}$$

with $\hat{G}_{y,i}(:, l)$ indicating the l-th column of $\hat{G}_{y,i}$ and the $|| \cdot ||_1$ norm measures absolute deviation.

To maintain consistency under varying noise levels, the coverage error is normalized as:

$$\tilde{e}(\hat{G}_{i,t}(k)) = \frac{e(\hat{G}_{i,t}(k))}{\max(1 \times 10^{-6}, ||y(k)||)}.$$

The penalty term is formulated as:

$$\begin{aligned}
\mathrm{Pen}(\hat{G}_{i,t}(k)) = &\, max(0, \hat{y}_{\mathrm{upper},i}(k) - y(k)) \\
&+ max(0, y(k) - \hat{y}_{\mathrm{lower},i}(k)),
\end{aligned}$$

where

$$\hat{y}_{\text{upper},i}(k) = \hat{p}_y(k) + \sum_{l=1}^{q} \left|\hat{G}_{y,i}(:,l)\right|,$$
$$\hat{y}_{\text{lower},i}(k) = \hat{p}_y(k) - \sum_{l=1}^{q} \left|\hat{G}_{y,i}(:,l)\right|.$$

Thus, the fitness evaluation for the i-th candidate is defined as:

$$F(\hat{G}_{i,t}(k)) = \lambda_1 \cdot \text{Vol}(\hat{G}_{i,t}(k)) + \lambda_2 \cdot \tilde{e}(\hat{G}_{i,t}(k)) + \lambda_3 \cdot \text{Pen}(\hat{G}_{i,t}(k)),$$

where λ_1, λ_2, and λ_3 are weighting coefficients. λ_3 is typically set large to enforce strict feasibility, while λ_1 and λ_2 can be tuned for balance. If an individual exceeds the feasible region after mutation and crossover, it is replaced randomly.

To improve diversity, mutation introduces new candidates. An elitist strategy ensures that the best individual $\hat{G}_{\text{best},t}(k)$ is preserved. The mutation scaling factor $\alpha_1(t)$ is adaptively defined as:

$$\begin{aligned} \alpha_1(t) = \alpha_{\min} + (\alpha_{\max} - \alpha_{\min}) \cdot \left[\exp\left(\frac{t}{T}\ln 2\right) - 1\right] \\ \cdot \left(\frac{1}{2} + \frac{F_{k,i} - F_{\min}}{F_{\max} - F_{\min}} \cdot \frac{1}{2}\right), \end{aligned} \tag{5.196}$$

where t is the generation index, T is the maximum evolutionary iterations, and $F_{k,i}$, $F_{\min}$, $F_{\max}$ denote the fitness of the individual, as well as the minimum and maximum fitness values, respectively. Another scaling factor $\alpha_2(t)$ is defined randomly in $[0, 1]$.

The mutation rule is given by:

$$\begin{aligned} \hat{G}'_{i,t}(k) = \hat{G}_{i,t}(k) + \alpha_1(t)(\hat{G}_{\text{best},t}(k) - \hat{G}_{i,t}(k)) \\ + \alpha_2(t)(\hat{G}_{r_1,t}(k) - \hat{G}_{r_2,t}(k)), \end{aligned} \tag{5.197}$$

where $\hat{G}_{r_1,t}(k)$ and $\hat{G}_{r_2,t}(k)$ are two randomly selected individuals.

Crossover then mixes information from $\hat{G}'_{i,t}(k)$ and $\hat{G}_{i,t}(k)$ to form the offspring,

$$\hat{G}^{\text{new}}_{i,k}(j,l) = \begin{cases} \hat{G}'_{i,k}(j,l), & \text{if } r_{j,l} \le CR, \\ \hat{G}_{i,k}(j,l), & \text{otherwise}, \end{cases} \tag{5.198}$$

where $r_{j,l}$ is a random number, and CR is the crossover probability. To enhance adaptability, CR is defined as:

$$\begin{aligned} CR = CR_{\min} + (CR_{\max} - CR_{\min})\left(2 - \exp\left(-\frac{t}{T}\ln 2\right)\right) \\ \cdot \left(\frac{1}{2} + \frac{F_{k,i} - F_{\min}}{F_{\max} - F_{\min}} \cdot \frac{1}{2}\right). \end{aligned} \tag{5.199}$$

By adaptively tuning $\alpha_1(t)$, $\alpha_2(t)$, and CR, the algorithm balances local exploitation in the early phase and global exploration in the subsequent phase, thereby accelerating convergence toward the global optimum. After each generation, the better individual between $\hat{G}_{i,k}^{\text{new}}$ and $\hat{G}_{i,t}(k)$ is retained.

5.6.3 Conclusion

The DE-based zonotopic feasible set filter (DE-ZKF) is proposed to mitigate the intrinsic conservativeness typically encountered in standard zonotopic Kalman filtering. The central innovation lies in embedding an adaptive optimization mechanism, whereby the generator matrix of the zonotope is continuously refined through a DE procedure. This evolutionary update dynamically reshapes the feasible set, leading to tighter state-bounding regions and a significant reduction of redundant directions.

The optimization process simultaneously accounts for multiple performance indices, such as the minimization of zonotope volume, the reduction of coverage error with respect to admissible trajectories, and the incorporation of penalty terms to enforce consistency. These objectives collectively guarantee that the estimated bounds remain both compact and resilient against bounded disturbances. Furthermore, the introduction of elitist preservation ensures that the best-performing candidate solutions are retained across generations, while adaptive adjustment of evolutionary parameters promotes a balanced search dynamic, combining global exploration with local exploitation. Such design features facilitate stable convergence toward near-optimal feasible sets without excessive computational burden.

In essence, the DE-ZKF framework enhances state estimation by providing enclosures that are simultaneously accurate, robust, and computationally tractable. By unifying zonotopic set representations with DE optimization, it establishes a systematic and generalizable approach for reliable estimation under bounded uncertainties, with potential applicability to high-dimensional, nonlinear, and time-varying systems.

5.7 Case Study on Battery Operating Status Analysis

To assess the applicability and performance of the newly introduced filtering and optimization methods in battery state-of-charge (SOC) estimation, this section conducts algorithm verification using a self-developed experimental platform for lithium-ion battery SOC analysis.

5.7.1 Case 1: PSO–OHSF

This case is based on the linear battery SOC estimation problem introduced in 5.1, where the cell dynamics are described by the Thevenin equivalent circuit model. The PSO–OHSF presented in 5.3 is applied to this system.

Based on the experimental platform described in 5.1, a series of charge–discharge tests were carried out to extract the OCV–SOC characteristics of the lithium-ion cell. The acquired data, summarized in Table 5.1, were then processed using MATLAB's `polyfit` function to perform polynomial regression, from which the following empirical OCV–SOC relationship was derived:

$$U_{\text{OCV}} = f(\text{SOC}) = 0.5293 \times \text{SOC} + 3.5821. \tag{5.200}$$

The state-space representation for SOC estimation is formulated as:

$$\begin{cases} \begin{bmatrix} \text{SOC}_{k+1} \\ U_{p,k+1} \end{bmatrix} = \begin{bmatrix} 1 & 0 \\ 0 & e^{-\frac{\Delta t}{R_p C_p}} \end{bmatrix} \begin{bmatrix} \text{SOC}_k \\ U_{p,k} \end{bmatrix} + \begin{bmatrix} -\dfrac{\eta \Delta t}{Q} \\ (1 - e^{-\frac{\Delta t}{R_p C_p}}) R_p \end{bmatrix} I_k + w_k, \\ U_k = 0.5293 \times \text{SOC} + 3.5821 - U_{p,k} - I_k R_0 + v_k, \end{cases} \tag{5.201}$$

where $R_0 = 0.0415\ \Omega$, $R_p = 0.3068\ \Omega$, $C_p = 2372.2\ F$, and $\tau_p = R_p C_p$ denotes the polarization time constant. The sampling interval is $\Delta t = 5\ s$, and the nominal capacity is $Q = 1.5$ Ah. The initial feasible set is $\mathcal{P}(\theta_{c,0}, T_0)$ with $\theta_{c,0} = [5, 1]^{\text{T}}$ and $T_0 = \text{diag}\{0.1, 0.1\}$. w_k and v_k are bounded as $w_{k,i} \in [-0.001, 0.001]$, $i = 1, 2$, and $v_k \in [-0.001, 0.001]$.

To emphasize the superiority of the presented PSO–OHSF algorithm, comparisons are made with the ellipsoidal set-membership filter (ESMF) [81] and the orthometric hyperparallel space set-membership filter (OHSSMF) [67]. The results are illustrated in Figures 5.6–5.8.

In Figure 5.6, the dotted boundary represents the feasible set enclosed by ESMF, the dashed boundary corresponds to OHSSMF, and the solid boundary denotes the proposed PSO–OHSF approach. The star marker indicates the true state. All three methods successfully enclose the true state and exhibit progressive contraction during system evolution, thereby demonstrating their effectiveness in maintaining bounded-error estimation. However, ESMF shows the highest degree of redundancy, which stems from the conservative nature of ellipsoidal bounds and

Table 5.1 OCV–SOC data table.

SOC/%	0	10	20	30	40	50	60	70	80	90	100
OCV/V	2.78	3.63	3.71	3.76	3.79	3.83	3.87	3.94	4.01	4.09	4.19

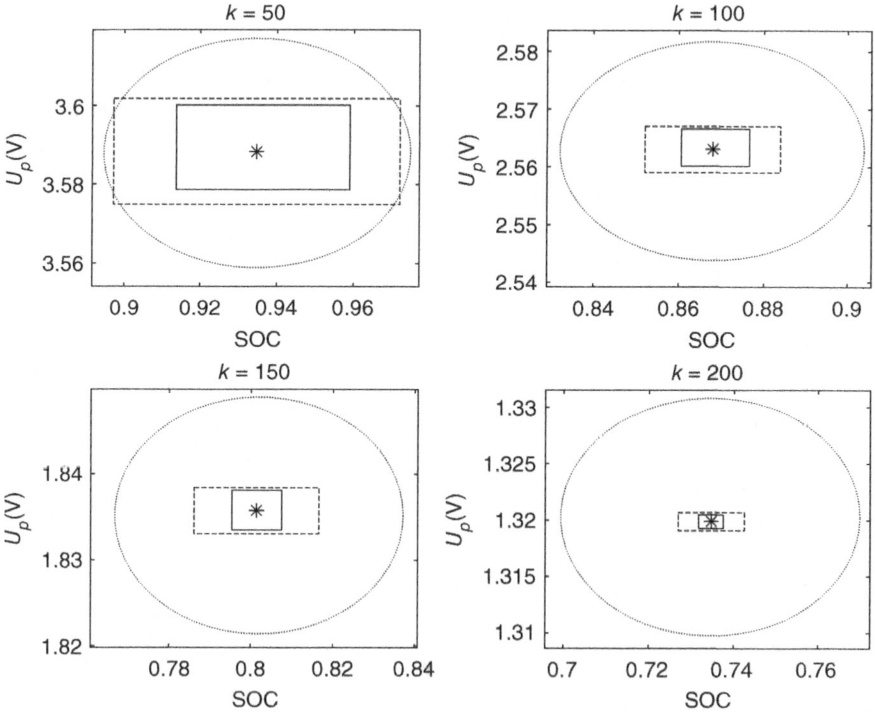

Figure 5.6 Comparison of feasible set domains produced by different algorithms at various time instants.

leads to looser feasible domains even at later iterations. By contrast, OHSSMF displays noticeable fluctuations at the early stages of iteration before achieving convergence. The proposed PSO–OHSF produces substantially more compact enclosures, reflecting reduced redundancy and enhanced conservativeness. This behavior confirms that swarm-assisted initialization and adaptive hyperparallel reconstruction enable PSO–OHSF to achieve both stability and efficiency in confining the feasible set close to the true state trajectory.

Figures 5.7 and 5.8 depict the estimated trajectories of SOC and U_p, together with their corresponding extremal bounds. It is clearly observed that the PSO–OHSF achieves faster convergence and higher accuracy in estimating both polarization voltage and SOC under UBB noise. In both figures, the magenta dashed lines illustrate that the ESMF generates overly broad bounds as a consequence of its ellipsoidal representation, which leads to excessive conservativeness and degraded estimation precision. The OHSSMF provides a certain improvement over ESMF by tightening the enclosures, but it still requires more iterations before reaching stable performance. In contrast, the PSO–OHSF, by embedding PSO

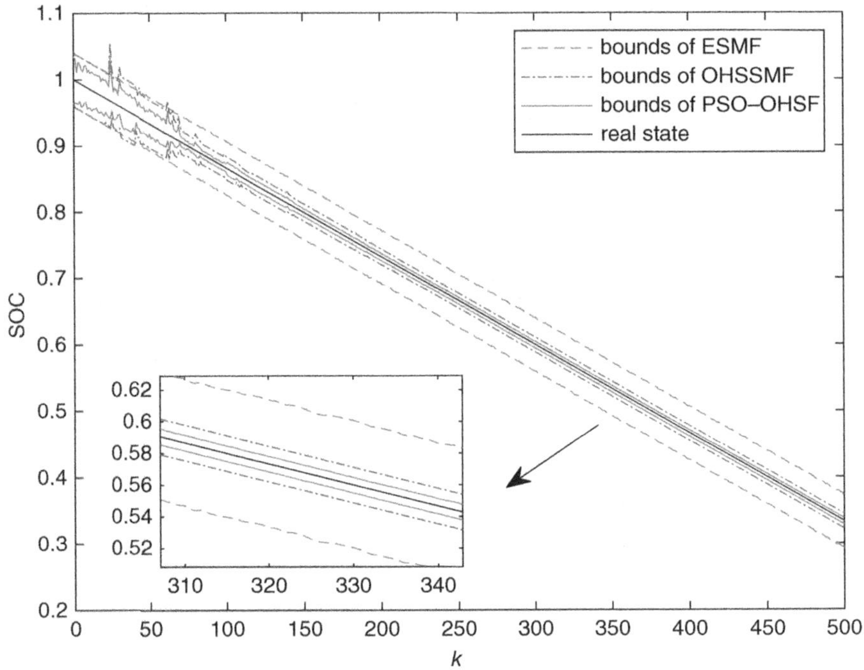

Figure 5.7 SOC state estimation comparison.

within the hyperparallel framework, constructs compact orthometric enclosures that significantly reduce conservativeness and yield feasible regions much closer to the true state trajectory. As a result, PSO–OHSF not only improves the estimation accuracy of SOC and U_p, but also demonstrates enhanced robustness and efficiency compared with the other two approaches.

5.7.2 Case 2: P-IPF

Based on the SOC model and the nonlinear system described in 5.1, the Improved particle filter algorithm based on the parallelotope introduced in 5.4 is applied. In this method, the parallelotope structure is used to approximate the feasible region, which enhances robustness against parameter variations.

In an environment with 25 °C, the fully charged battery was discharged at a constant current of 0.5C. The obtained experimental data is shown in Figure 5.9. According to the OCV–SOC correspondence recorded every 10% SOC (Table 5.2), parameters are identified using least squares as $R_0 = 0.0415\ \Omega$, $R_p = 0.03068\ \Omega$, and $C_p = 2372.2$ F.

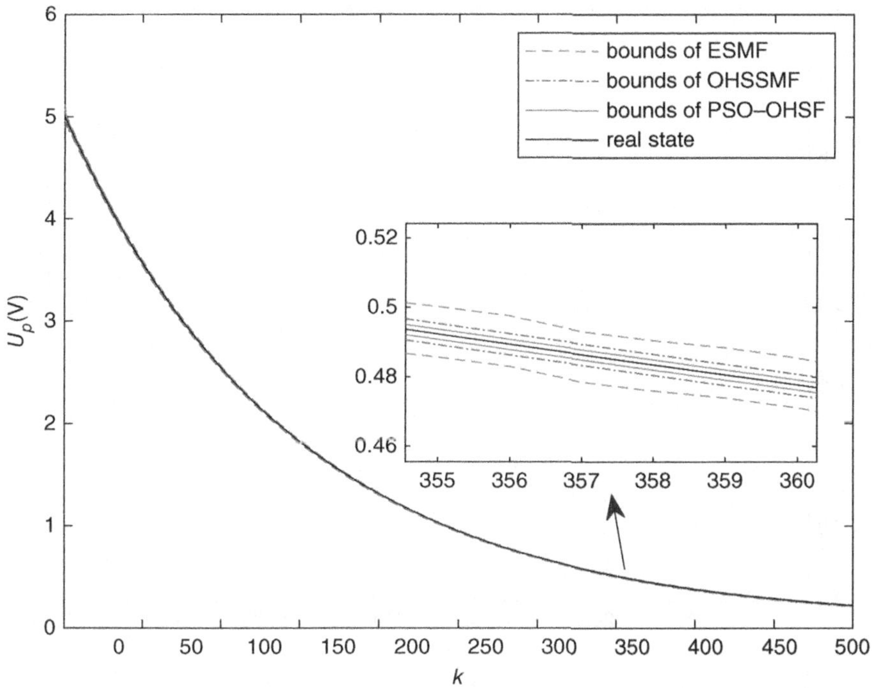

Figure 5.8 U_p state estimation comparison.

State Estimation: Based on the OCV–SOC data in Table 5.2, a cubic polynomial is fitted, leading to the nonlinear measurement model:

$$\begin{aligned} U_k = {} & 5.83\,\mathrm{SOC}_k^3 - 9.27\,\mathrm{SOC}_k^2 + 5.20\,\mathrm{SOC}_k + 2.943 \\ & -U_{p,k} - I_k R_0 + \upsilon_k. \end{aligned} \tag{5.202}$$

The system is initialized with the state vector $x_0 = [1 \;\; 0.1]^{\mathrm{T}}$, with feasible set $\mathcal{P}(\hat{x}_0, G_0)$ where $\hat{x}_0 = [1 \;\; 0.1]^{\mathrm{T}}$ and $G_0 = \mathrm{diag}\{0.03, 0.01\}$. The sampling period is $\Delta t = 5$ s. w_k and υ_k are both bounded as $|w_{k,i}| \leq 0.01$ $(i = 1, 2)$ and $|\upsilon_k| \leq 0.01$.

The root-mean-square error (RMSE) is adopted:

$$\mathrm{RMSE} = \sqrt{\frac{1}{S} \sum_{s=1}^{S} \left(x_k^s - \hat{x}_k^s\right)^2}, \tag{5.203}$$

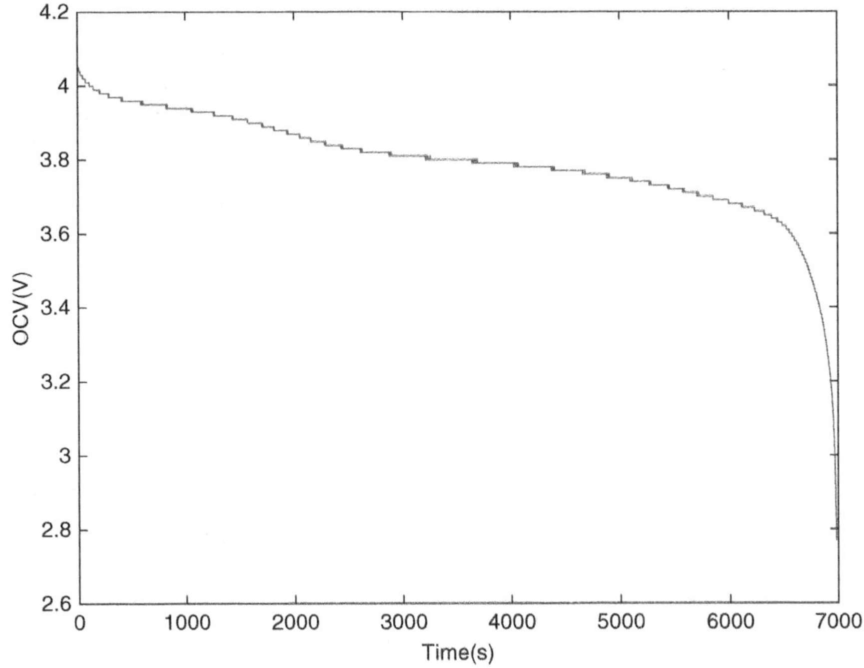

Figure 5.9 OCV curve.

Table 5.2 OCV–SOC correspondence data.

SOC(%)	0	10	20	30	40	50	60	70	80	90	100
OCV(V)	2.78	3.63	3.71	3.76	3.79	3.83	3.87	3.94	4.01	4.09	4.19

where S denotes the number of Monte Carlo runs. The non-Gaussian noise follows the mixed distribution [82]:

$$\begin{aligned} \mathrm{PDF}_{\text{noise}} &= \lambda \cdot \frac{1}{\sqrt{2\pi\sigma^2}} \exp\left(-\frac{(x-\mu)^2}{2\sigma^2}\right) \\ &\quad + (1-\lambda) \cdot \frac{1}{b-a}, \quad x \in [a, b], \end{aligned} \tag{5.204}$$

with $\lambda = 0.5$, variance $\sigma^2 = 0.0001$, and uniform bounds $a = -0.01$, $b = 0.01$.

Figure 5.10 shows the feasible domains obtained by extended set membership particle filter (ESMPF) [83] and P-PF. It indicates that both approaches successfully enclose the true states, as marked by the star markers, thereby validating

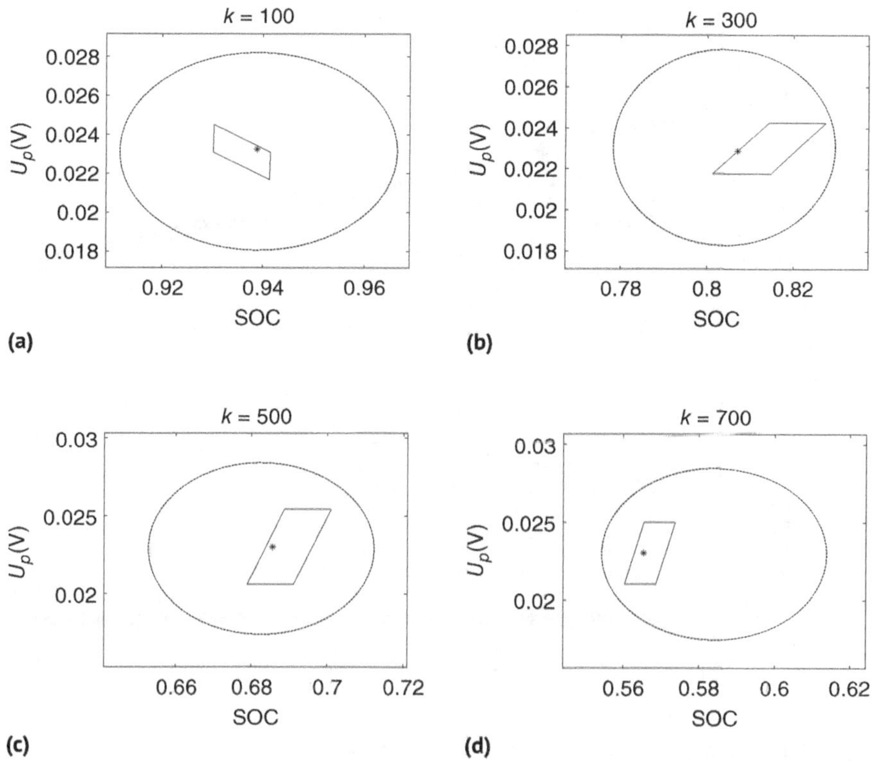

Figure 5.10 Comparison of feasible particle domains of ESMPF and P-IPF. Star markers denote true states; dashed ellipses indicate ellipsoidal feasible regions (ESMPF); solid quadrilaterals indicate parallelotope feasible regions (P-IPF).

their effectiveness in bounding the estimation error. However, the ellipsoidal feasible regions produced by ESMPF are relatively conservative and lead to redundant enclosures, whereas the parallelotope representation adopted in P-IPF generates a more compact domain. This tighter bounding not only reduces conservatism but also improves the fidelity of the state approximation, which translates into higher estimation accuracy and enhanced robustness against parameter variations.

In Figure 5.11, the solid line corresponds to the true state obtained from the constant-current discharge experiment. The EKF algorithm provides satisfactory filtering performance under the Gaussian noise assumption. However, when the noise is UBB, EKF fails to accurately estimate the state, as illustrated by the solid line with diamond markers. In contrast, the maximum correlation entropy criterion adaptive extended Kalman filter (MCC-AEKF) algorithm [82] achieves more accurate estimation in the presence of non-Gaussian noise, as indicated by the dashed line with circle markers. The dashed line with triangle markers corresponds to the extended parallelotope set membership filter

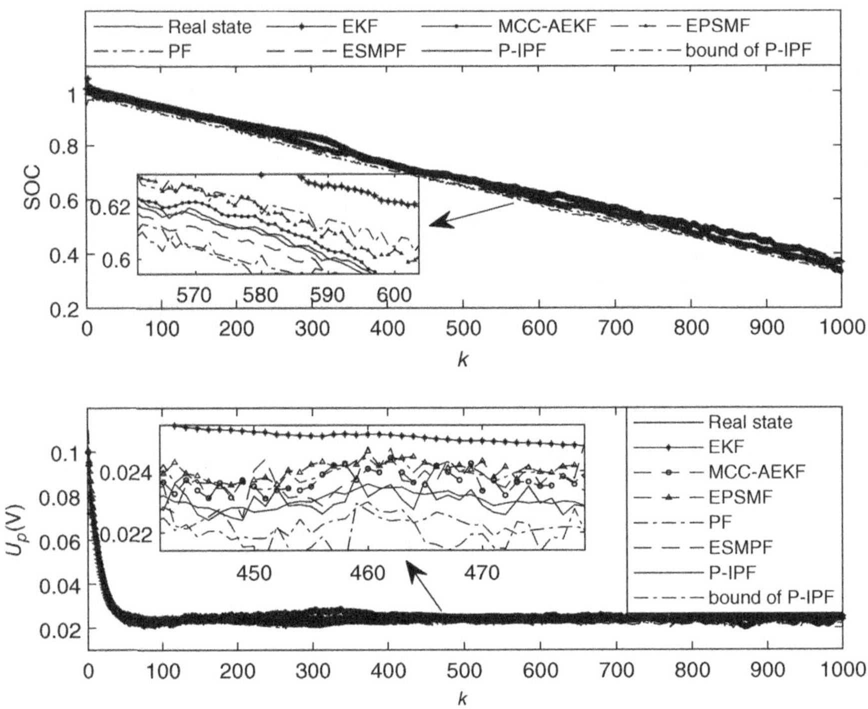

Figure 5.11 Comparison of state estimation results.

(EPSMF) algorithm [62], which estimates the state by constructing a parallelotope space that encloses the true trajectory, with the center of this space taken as the estimate. Although this approach yields better performance than EKF and PF in the UBB noise case, its accuracy remains slightly lower than that of MCC-AEKF. As summarized in Table 5.3, both EKF and PF achieve lower computation times but suffer from larger RMSE values. Among the PF-based methods with $N = 20$, the ESMPF algorithm improves robustness by employing an ellipsoidal feasible region to mitigate particle degradation. Compared with EPSMF, ESMPF increases computation time by 0.44 s but reduces the RMSE of SOC and U_p by 34.7% and 25.1%, respectively. Its RMSE is comparable to MCC-AEKF, though at the cost of an additional 1.01 s runtime. The proposed P-IPF algorithm further enhances estimation accuracy: while its computation time rises by 0.29 s relative to MCC-AEKF, the RMSE of SOC and U_p are reduced by 36.4% and 26.9%, respectively. These results demonstrate that the P-IPF algorithm effectively balances computational cost and estimation accuracy, providing a clear improvement over existing approaches.

The bold values indicate the best performance for each evaluation metric. Specifically, the bold values "0.00319" and "0.000703" denote the smallest estimation errors achieved by the P-IPF algorithm. The bold value "5.47" denotes the shortest computation time achieved by the EKF algorithm.

As illustrated in Figure 5.12, the feasible domain area rapidly converges within approximately eight iterations, indicating that the parallelotope representation stabilizes after only a few updates. In addition, the dash-dotted lines in Figure 5.11 show that the confidence region consistently encloses the true state throughout the estimation horizon, which demonstrates that P-IPF is capable of maintaining particles within a bounded feasible set of nearly constant scale. This behavior highlights that the algorithm not only ensures reliable state containment but also

Table 5.3 Comparison of RMSE and computation time of six algorithms.

	EKF	MCC-AEKF	EPSMF	PF	ESMPF	P-IPF
RMSE of SOC	0.0175	0.00502	0.00736	0.00879	0.00480	**0.00319**
RMSE of U_p	0.00264	0.000962	0.00135	0.00169	0.00101	**0.000703**
Running time (s)	**5.47**	7.36	8.13	6.21	8.57	7.65

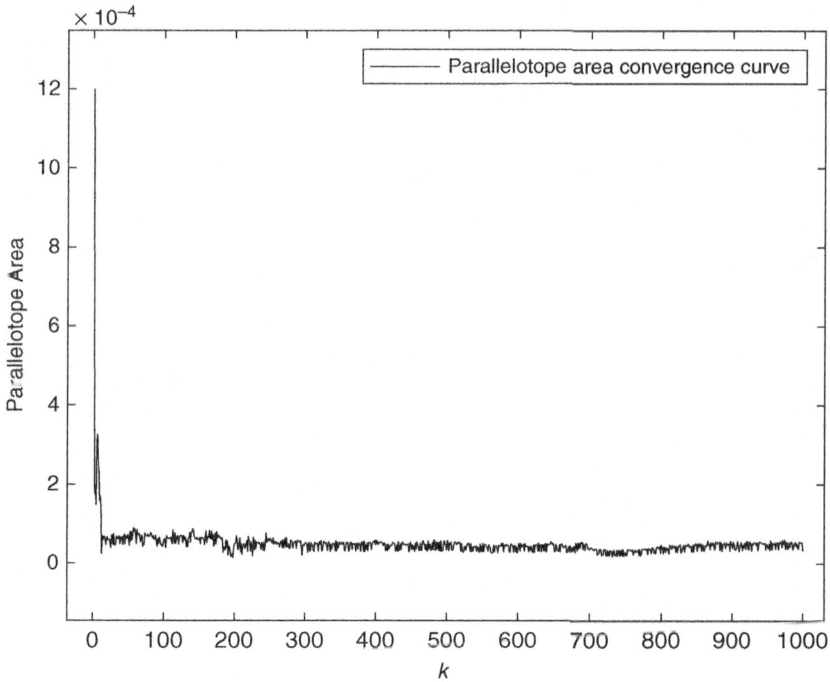

Figure 5.12 Convergence of parallelotope area.

effectively suppresses particle dispersion, thereby improving both stability and estimation accuracy in the presence of bounded uncertainties.

Analysis of the Influence of SOC Initial Value on Estimation Accuracy: The correctness of the initial battery SOC value has a significant impact on determining the precision of the estimation results. For evaluating the algorithm's robustness against uncertainty or errors in the initial SOC specification, two reference cases are considered: one with an accurate initial SOC of 1.0, and another with an inaccurate initial SOC of 0.7. The estimation process is performed under an environment characterized by UBB noise.

The corresponding estimation trajectories and error curves are presented in Figure 5.13. It can be observed that, despite the presence of a substantial initial SOC deviation, the P-IPF algorithm quickly drives the estimates toward the true value once the battery operation commences. This behavior demonstrates that P-IPF is capable of effectively correcting initialization errors, thereby ensuring robustness against uncertain conditions and maintaining high accuracy throughout the estimation horizon.

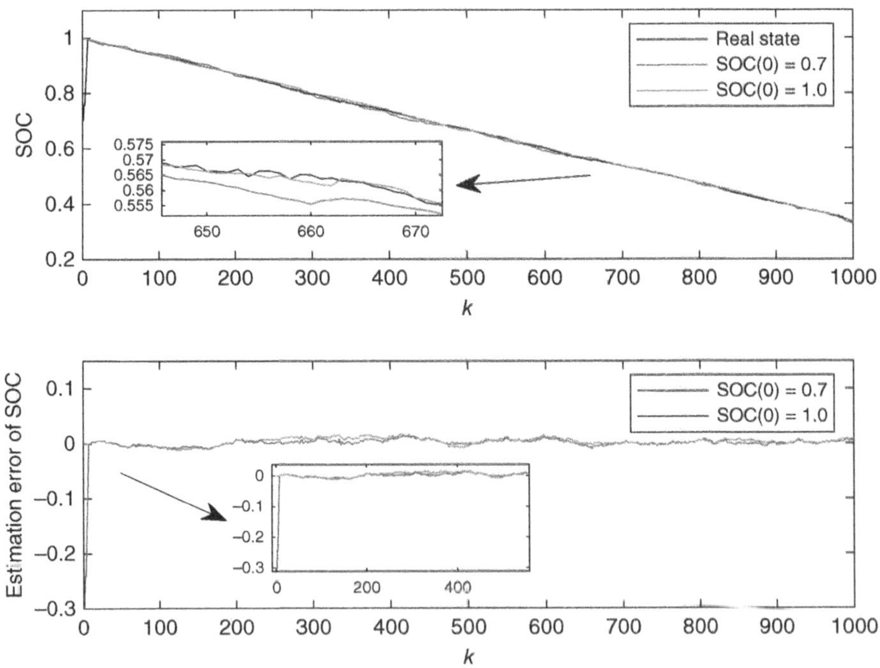

Figure 5.13 Estimation results and errors of the P-IPF algorithm under different initial SOC values.

5.7.3 Case 3: PPC-ZSF

The case focuses on the application of the PPC-ZSF presented in 5.5. This approach augments the basic zonotopic filtering scheme by introducing confinement and projection mechanisms, which effectively suppress particle divergence and enforce tighter bounds on the SOC trajectory.

A fully charged lithium-ion battery was subjected to constant-current discharge, and the OCV was recorded at every 10% decrement in SOC. The sampled OCV and SOC values are shown in Table 5.4. As summarized in Table 5.4, OCV remains relatively stable within the SOC range of 10–90%. To quantitatively describe this relationship, MATLAB's `polyfit` function was used to perform a first-order linear regression, yielding the empirical expression

$$U_{\text{OCV}} = f(\text{SOC}) = 0.5293 \times \text{SOC} + 3.5821. \tag{5.205}$$

The state-space equation that characterizes the relationship between SOC_k and U_k can be expressed as:

$$\begin{cases} \begin{bmatrix} \text{SOC}_{k+1} \\ U_{p,k+1} \end{bmatrix} = \begin{bmatrix} 1 & 0 \\ 0 & e^{-\frac{\Delta t}{R_p C_p}} \end{bmatrix} \begin{bmatrix} \text{SOC}_k \\ U_{p,k} \end{bmatrix} + \begin{bmatrix} -\dfrac{\eta \Delta t}{Q} \\ (1 - e^{-\frac{\Delta t}{R_p C_p}}) R_p \end{bmatrix} I_k + w_k, \\ U_k = 0.5293 \times \text{SOC}_k + 3.5821 - U_{p,k+1} - I_{k+1} R_0 + v_k, \end{cases} \tag{5.206}$$

where the circuit parameters are summarized in Table 5.5.

Table 5.4 Corresponding relationship between OCV and SOC.

SOC (%)	OCV (V)	SOC (%)	OCV (V)
0	2.782	60	3.873
10	3.629	70	3.938
20	3.711	80	4.012
30	3.758	90	4.088
40	3.792	100	4.193
50	3.826		

Table 5.5 Parameters of the first-order Thevenin equivalent circuit.

R_0 (Ω)	R_p (Ω)	I (A)	C_p (F)	Δt (s)	Q_{real} (Ah)	η
0.0421	0.03112	1	2369.8	5	1200	0.9

Based on prior knowledge of battery characteristics and historical data, the system initial state together with process and measurement noise are represented by the following zonotopes:

$$\begin{bmatrix} \mathrm{SOC}_0 \\ U_{p,0} \end{bmatrix} \in \left\langle \begin{bmatrix} 1 \\ 5 \end{bmatrix}, \begin{bmatrix} 0.005 & 0 \\ 0 & 0.005 \end{bmatrix} \right\rangle, \tag{5.207}$$

$$w_k \in \left\langle \begin{bmatrix} 0 \\ 0 \end{bmatrix}, \begin{bmatrix} 0.001 & 0 \\ 0 & 0.001 \end{bmatrix} \right\rangle, \quad v_k \in (0, 0.001). \tag{5.208}$$

For performance validation, the SOC estimation results of the PPC-ZSF approach are contrasted with those obtained by the ESMF [84] and the zonotopic set-membership filter (ZSMF) [85].

Figure 5.14 shows the feasible state regions generated by ESMF, ZSMF, and the proposed PPC-ZSF at different time instants. All three methods ensure that the actual state remains within the corresponding feasible region and that the regions are consistently updated as the system evolves. Nevertheless, the ellipsoidal domains produced by ESMF are considerably large, reflecting substantial

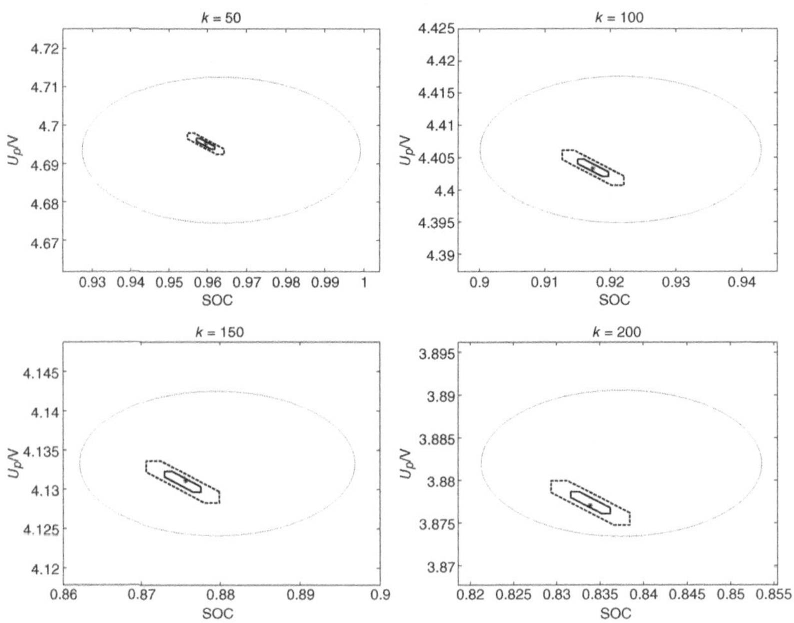

Figure 5.14 Comparison of SOC real state and feasible estimation sets for different algorithms. The dotted, dashed, and solid contours represent feasible sets derived by ESMF, ZSMF, and the proposed PPC-ZSF, respectively, at $k = 50, 100, 150, 200$.

redundancy and insufficient compactness. The ZSMF alleviates this drawback by providing more compact enclosures with reduced redundancy, yet the resulting feasible sets are still relatively expansive. In contrast, the PPC-ZSF consistently delivers smaller and more compact feasible regions across iterations, tightly enclosing the true state trajectory. This property not only minimizes redundancy but also enhances the fidelity of the state approximation, leading to higher estimation accuracy and improved robustness against bounded disturbances.

Figures 5.15 and 5.16 display the comparison of estimation bounds for SOC and polarization voltage U_p, respectively. As shown, the feasible regions generated by all three algorithms successfully encapsulate the true trajectories of the estimated variables, ensuring bounded-error consistency. Nevertheless, the ESMF method produces the widest bounds and exhibits pronounced fluctuations during periods of rapid state variation, suggesting that it imposes weaker restrictions on the feasible search space under dynamic UBB noise and thereby compromises estimation precision. The ZSMF algorithm alleviates part of this limitation by providing moderately tighter bounds, yet its conservative adjustment strategy leads to broader ranges during fast transitions and prevents full utilization of system dynamics. In contrast, the proposed PPC-ZSF consistently produces narrower and more compact feasible sets that closely adhere to the true state evolution. This improved

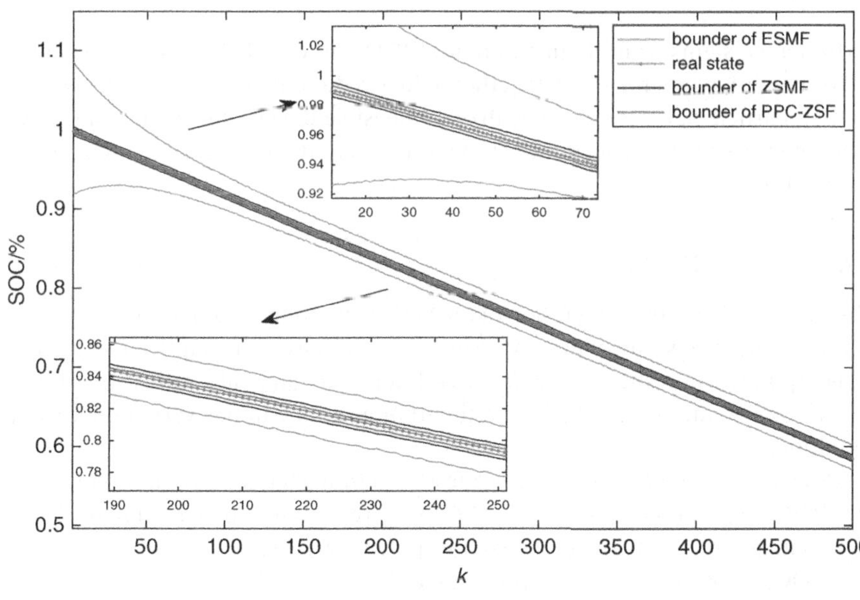

Figure 5.15 Comparison of SOC's real state and estimation bounds for different algorithms.

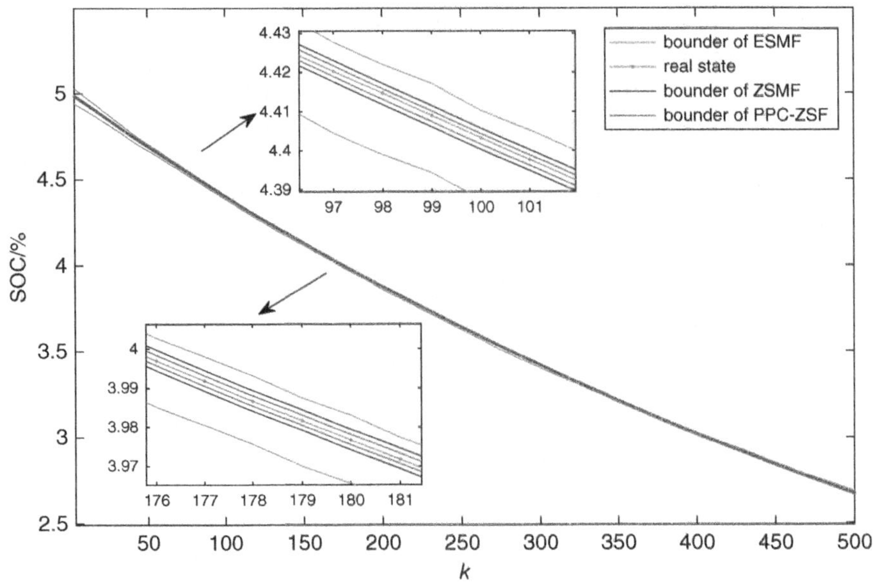

Figure 5.16 Comparison of U_p's real state and estimation bounds for different algorithms.

performance stems from its integration of PSO-based global search with spatial constraint enforcement, which together reduce redundancy in the search space by guiding particle behavior and penalizing infeasible deviations. As a result, PPC-ZSF achieves stable and accurate estimation even under strongly dynamic operating conditions.

5.7.4 Case 4: DE-ZKF

This case examines the integration of the SOC estimation model from 5.1 with the advanced filtering strategy presented in 5.6. In this case, an adaptive mechanism is incorporated into the zonotopic space filter to dynamically adjust the uncertainty bounds, aiming to improve estimation accuracy under varying operating conditions.

Using a constant-current discharge test on a fully charged lithium-ion battery, the OCV was recorded at intervals of 10% SOC reduction. The resulting measurement data are presented in Table 5.6. By applying MATLAB's `polyfit` function to the OCV–SOC data, a linear approximation is obtained:

$$U_{\text{OCV}} = f(\text{SOC}) = 1.0798 \cdot \text{SOC} + 3.0657, \tag{5.209}$$

where $f(\cdot)$ denotes the regression function. Accordingly, the discrete-time model derived from (5.2) is expressed as (5.3).

where U_p, k and U_k denote the polarization voltage and terminal voltage, respectively; $\tau_p = R_p C_p$ is the polarization time constant, and $T_s = 0.9$ s is the sampling period. The system is initialized with $p(0) = [1 \quad 5]^{\mathrm{T}}$ with covariance matrix $G(0) = 0.1^2 I_2$. The process noise is bounded by $\| w(k) \|_\infty \leq 10^{-3}$, while the measurement noise is constrained by $\| v(k) \|_\infty \leq 10^{-3}$. The number of steps is $N = 20$ with a time horizon of $T = 20$. The parameters characterizing the experimental platform are provided in Table 5.7.

Table 5.6 Relationship between OCV and SOC under experimental conditions.

SOC (%)	OCV (V)	SOC (%)	OCV (V)
0	2.70	55	3.76
5	2.89	60	3.78
10	3.03	65	3.80
15	3.17	70	3.83
20	3.27	75	3.86
25	3.45	80	3.89
30	3.52	85	3.91
35	3.58	90	3.93
40	3.60	95	3.96
45	3.65	100	4.11
50	3.74		

Table 5.7 Battery parameters.

Parameter	Value
I_c	1 A
R_0	0.0415 Ω
R_p	0.03068 Ω
C_p	480.3 F
Q	205 Ah
η	0.6

To highlight the behavior of the proposed DE-ZKF, it is benchmarked against ESMF [86], ZSMF [87], and the standard ZKF [88]. The results are illustrated in Figures 5.17–5.19.

Figures 5.17 and 5.18 present the evolution of SOC and U_p together with estimation bounds obtained from four different algorithms. Among them, the ESMF method generates the widest intervals, which reflects its inherent conservativeness and results in looser state enclosures. The ZSMF and ZKF algorithms provide moderately improved performance with somewhat tighter bounds, yet they still display noticeable conservatism, particularly during dynamic variations. In contrast, the DE-ZKF demonstrates rapid convergence toward the true state and consistently produces significantly narrower bounds. This behavior indicates that DE-ZKF effectively reshapes the feasible sets to achieve higher compactness while

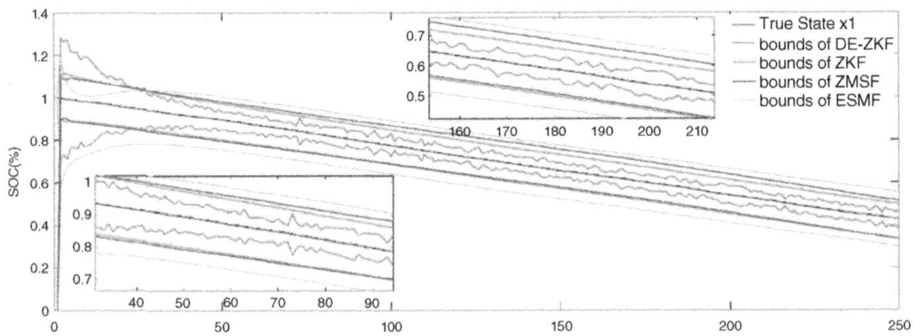

Figure 5.17 Comparison of SOC bounds at different times.

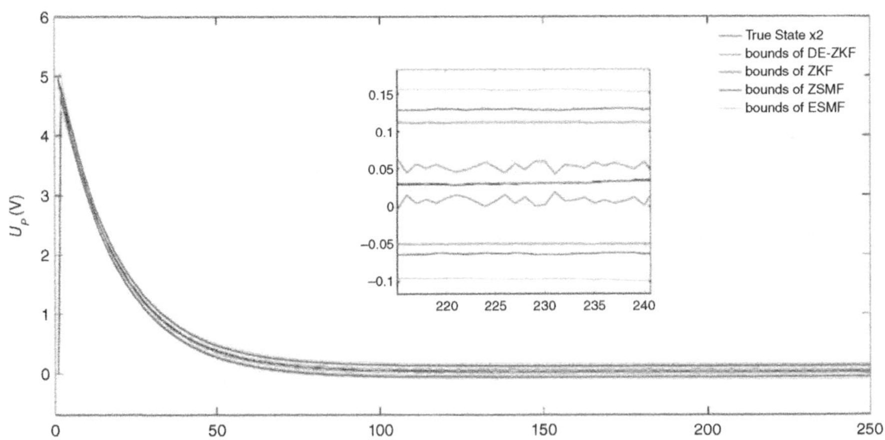

Figure 5.18 Comparison of U_p bounds at different times.

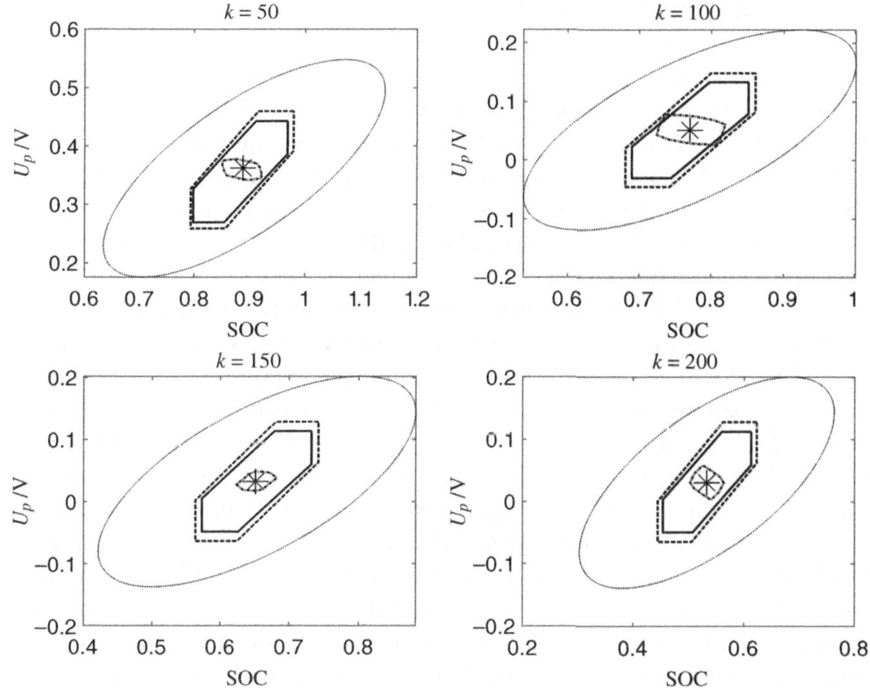

Figure 5.19 Comparison of feasible set spaces obtained by four algorithms at different times.

maintaining accurate tracking of the state trajectory, thereby combining robustness with improved estimation precision.

Figure 5.19 provides a detailed comparison of feasible sets at several representative time instants. The solid curve indicates the true state trajectory, while the dotted, dashed, solid, and dash-dotted contours correspond to ESMF, ZSMF, ZKF, and DE-ZKF, respectively. It is evident that all four approaches successfully enclose the true state within their feasible regions, confirming their validity under bounded uncertainties. However, the sets produced by ESMF, ZSMF, and ZKF remain relatively large and exhibit varying degrees of spatial redundancy, reflecting conservative approximations of the admissible domain. In contrast, the DE-ZKF consistently produces the smallest feasible volumes across iterations, thereby reducing redundancy and achieving more compact state enclosures. This improvement highlights its ability to provide higher estimation precision and enhanced robustness in dynamic operating conditions.

5.8 Concluding Remarks

This chapter has focused on the problem of lithium-ion battery SOC estimation from the standpoint of set-membership filtering and its algorithmic extensions. The state dynamics were represented by a Thevenin-type equivalent circuit model, and the OCV–SOC curve was experimentally identified through constant-current charge–discharge tests. On this foundation, the SOC estimation task was reformulated within a zonotopic framework, allowing the feasible set of states to be recursively propagated through prediction and correction steps.

Several advanced filtering strategies were proposed and comparatively analyzed. The PSO-based orthometric hyperparallel space filter (PSO–OHSF) improved the initialization of the feasible region and mitigated search-space expansion by employing a geometrically regular orthometric construction. The parallelotope-based improved particle filter (P-IPF) enhanced resilience to parameter variations by bounding the admissible state trajectories with a minimal-volume parallelotope, thereby maintaining reliable state enclosures. The projected particle-confinement zonotopic space filter (PPC-ZSF) effectively counteracted particle divergence and ensured tighter SOC bounds under rapid transient conditions. Finally, the DE-based zonotopic Kalman filter (DE-ZKF) adaptively reshaped generator matrices and tuned uncertainty bounds via evolutionary optimization, striking a balance between compactness and robustness in the estimation process.

Extensive simulations validated that these methods consistently yield accurate and reliable SOC estimation while offering distinct trade-offs among computational burden, estimation precision, and robustness against uncertainty. In particular, swarm-assisted strategies such as PSO–OHSF and projection-driven approaches like PPC-ZSF provided markedly tighter bounds without compromising stability, whereas geometry-based schemes such as P-IPF and optimization-driven designs like DE-ZKF demonstrated strong robustness and adaptability. Overall, the findings highlight that integrating physics-informed battery modeling with set-membership filtering and intelligent metaheuristic optimization constitutes a powerful and practical framework for achieving accurate, robust, and efficient SOC estimation in real-world battery management systems.

6

Advances in Battery Remaining Useful Life Analysis by Filtering and Artificial Intelligence

6.1 Introduction to Battery Remaining Useful Life Analysis

6.1.1 Definition and Significance of Remaining Useful Life

Lithium-ion batteries, serving as the cornerstone of modern energy storage systems, are widely deployed in a spectrum of applications ranging from electric vehicles and portable electronics to renewable energy integration and aerospace systems, owing to their high energy density, long cycle life, and low self-discharge rate. However, these electrochemical devices are subject to intricate and irreversible degradation mechanisms during repeated charge–discharge cycles. Internal processes such as the continuous growth and reconstruction of the solid electrolyte interphase (SEI) film, loss of active lithium inventory, and structural degradation of electrode materials collectively contribute to a gradual yet persistent deterioration of battery performance [89, 90].

Within the framework of prognostics and health management (PHM), the concept of remaining useful life (RUL) has emerged as a pivotal metric. RUL is quantitatively defined as the number of remaining charge–discharge cycles or the duration of service time from the current moment until the battery's performance degrades to a predefined end-of-life (EOL) threshold, beyond which it can no longer meet the operational requirements of the specific application [89]. A widely adopted failure criterion is the point at which the battery's actual capacity fades to 70–80% of its nominal rated capacity [91].

Accurate prediction of RUL is of paramount importance for ensuring:

- Safety: It enables the early warning of potential battery failures, such as thermal runaway, thereby preventing catastrophic accidents and enhancing operational safety, particularly in critical systems like electric vehicles and aircraft.

- Reliability: It provides a quantitative basis for maintenance scheduling, replacement decisions, and system redundancy management, ensuring the continuous and reliable operation of battery-powered systems.
- Economic Efficiency: It facilitates the implementation of predictive maintenance strategies, minimizing unscheduled downtime. Furthermore, precise RUL assessment is fundamental for enabling the second-life use of retired batteries in less demanding applications, thus maximizing their lifecycle value and promoting a circular economy.

6.1.2 Challenges and Current Research Limitations in RUL Prediction

Despite its critical importance, achieving accurate and robust RUL prediction remains a formidable challenge due to the inherent complexities of battery degradation and practical operational constraints. The primary difficulties and corresponding limitations in current research are summarized as follows:

1. Highly Nonlinear and Time-varying Degradation Trajectory: Battery capacity decay often exhibits complex nonlinear behavior, which may not follow a fixed pattern. The degradation rate can dynamically change over cycles, potentially showing phases of rapid fade followed by slower decay or even sudden "knee-point" behavior. This strong nonlinearity poses significant challenges to the generalization capability of prediction models, as simplistic linear or single-mode empirical models often fail to capture the entire degradation process accurately.
2. Complex Internal States and "Black-Box" Characteristics: The internal states of a battery (e.g., lithium-ion concentration, side reaction rates) are not directly measurable and must be inferred indirectly from external parameters such as voltage, current, and temperature. This "gray-box" or "black-box" nature makes precise modeling based on first principles extremely difficult. Moreover, model parameters themselves often evolve with aging, complicating the task of state tracking and parameter identification [92].
3. Capacity Regeneration Phenomena and Noise Interference: A notable challenge is the occurrence of capacity regeneration, where the measured capacity exhibits temporary, slight recoveries during long-term cycling, attributed to temporary redistributions of ions or relaxation effects. This phenomenon, coupled with measurement noise and fluctuations in operational conditions, severely obscures the underlying long-term degradation trend and can easily lead to erroneous predictions if not properly handled [93].
4. Dual Uncertainty in Models and Data: RUL prediction is inherently an exercise in managing uncertainty. This encompasses: (i) Process noise reflecting

model inaccuracies; (ii) Measurement noise from data acquisition systems; and (iii) Parameter uncertainty, including unknown initial states and noise statistics. Traditional stochastic filters often assume Gaussian noise distributions, an assumption that frequently fails in real-world complex environments, leading to biased estimates and overconfident predictions.

5. Computational Efficiency Versus Real-time Requirements: High-fidelity models or sophisticated algorithms are often computationally intensive, making them unsuitable for real-time online prediction on resource-constrained hardware like embedded battery management systems (BMS). Striking a balance between prediction accuracy and computational tractability is a key bottleneck for practical deployment [94].

To address these challenges, the subsequent sections of this chapter will introduce innovative solutions that synergistically combine advanced filtering techniques and artificial intelligence (AI) strategies. These approaches aim to overcome the current limitations in RUL prediction by incorporating more robust noise handling mechanisms, enhancing algorithmic efficiency through dynamic complexity reduction (DCR) and improved resampling strategies, and leveraging hybrid frameworks that integrate signal processing with deep learning.

6.1.3 Principal Technical Routes and Chapter Focus

To address the aforementioned challenges, the research community has developed a variety of RUL prediction methodologies, which can be broadly categorized into three principal technical routes, each with distinct characteristics and applicability.

1. Model-based Approaches: These approaches rely on mathematical models that describe the battery's degradation dynamics, encompassing physics-based electrochemical models, equivalent circuit models, or empirical decay models, such as the double exponential model. Algorithms such as the Kalman filter and its variants, including extended Kalman filter, unscented Kalman filter, or the particle filter, are employed to recursively update the model states and parameters based on incoming measurement data. The updated model is then extrapolated to the failure threshold to estimate the RUL [89, 90]. The primary strength of model-based methods lies in their incorporation of physical insights, providing interpretable predictions. However, their accuracy is contingent upon the fidelity of the underlying model, and they often face a trade-off between model complexity and computational feasibility for real-time applications.
2. Data-driven Approaches: Data-driven methods circumvent the need for explicit physical or empirical models by treating the battery as a "black-box."

They directly learn the mapping from historical operational data, such as voltage, current, temperature, and capacity sequences, to the degradation trend and RUL. With their powerful capability for automatically extracting temporal features and modeling complex nonlinearities, AI techniques, particularly deep learning models like the long short-term memory (LSTM) network, have become predominant in this category [95, 96]. While these methods often demonstrate high accuracy and adaptability to diverse operating conditions, they typically require substantial amounts of high-quality training data and may lack physical interpretability.

3. Hybrid and Fusion Approaches: Recognizing the complementary strengths and limitations of model-based and data-driven methods, hybrid or fusion strategies have emerged as a promising direction. These approaches aim to synergistically combine the prior knowledge embedded in mechanistic or empirical models with the adaptive learning capacity of data-driven models. Prominent fusion paradigms include coupling filtering algorithms with neural networks for joint state estimation and prediction, employing signal processing techniques, such as ensemble empirical mode decomposition (EEMD), to preprocess raw data and enhance input quality for data-driven models, and using optimization algorithms to fine-tune model parameters [97, 98]. The core objective is to achieve more robust, accurate, and generalizable RUL predictions by leveraging the best of both worlds.

Building upon this technical landscape, this chapter focuses specifically on the forefront of RUL prediction research, where advanced filtering techniques and AI strategies are being synergistically integrated. The subsequent sections will delve into innovative methodologies that exemplify this trend, demonstrating how such integrations effectively enhance prediction accuracy, robustness, and computational efficiency, thereby addressing the critical challenges outlined in Section 6.1.2.

6.2 Dynamic Complexity Reduction Zonotopic Kalman Filter

In this section, a dynamic complexity reduction zonotopic Kalman filter (DCR-ZKF) is developed for the joint estimation of state-of-charge (SOC) and state-of-health (SOH) in lithium-ion batteries. The key innovation of the proposed approach lies in its ability to dynamically optimize the zonotopic representations associated with voltage and current, thereby enhancing both the robustness and accuracy of the estimation compared with conventional zonotopic filtering methods.

6.2.1 Preliminaries and Problem Formulation

6.2.1.1 Preliminaries

This section summarizes the notation and fundamental concepts used throughout the chapter. The symbol $\mathbb{R}$ denotes the set of real numbers, and I_m represents the m-dimensional identity matrix. The set $B^m = [-1, 1]^m$ denotes the unit hypercube, while the Minkowski sum of sets $\mathcal{X}$ and $\mathcal{Y}$ is defined as $\mathcal{X} \oplus \mathcal{Y} = \{x + y : x \in \mathcal{X},\ y \in \mathcal{Y}\}$. The operator $\mathrm{tr}(\cdot)$ stands for the matrix trace. Several essential definitions, properties, and lemmas are recalled below to support the subsequent developments.

Definition 6.1 An n-dimensional zonotope of order m is any set $\mathcal{Z} \subset \mathbb{R}^n$ that can be written as the affine image of the unit hypercube B^m (with $m \geq n$) [99], namely

$$\mathcal{Z} = p \oplus HB^m = \{\, p + Hz : z \in B^m \,\} \triangleq \langle p, H \rangle, \tag{6.1}$$

where $p \in \mathbb{R}^n$ is referred to as the center of $\mathcal{Z}$ and $H \in \mathbb{R}^{n\times m}$ is its generator (shape) matrix.

Property 8 Let $\mathcal{Z}_1 = \langle p_1, H_1 \rangle \subset \mathbb{R}^n$ and $\mathcal{Z}_2 = \langle p_2, H_2 \rangle \subset \mathbb{R}^n$ be two zonotopes. Their Minkowski sum is again a zonotope [99], given explicitly by

$$\mathcal{Z}_1 \oplus \mathcal{Z}_2 = \left\langle p_1 + p_2,\ [\, H_1 \ \ H_2 \,] \right\rangle. \tag{6.2}$$

Property 9 Consider a zonotope $\mathcal{Z} = \langle p, H \rangle \subset \mathbb{R}^n$ and a matrix $K \in \mathbb{R}^{\ell\times n}$. The image of $\mathcal{Z}$ under the linear map $x \mapsto Kx$ is still a zonotope [99], denoted by

$$K \odot \mathcal{Z} = \langle Kp,\ KH \rangle. \tag{6.3}$$

Lemma 6.1 *Consider a zonotope $\mathcal{Z} = \langle p, H \rangle \subset \mathbb{R}^n$. The operator $\downarrow_s H$ performs a complexity reduction on the generator matrix H, producing a lower-order representation such that $\langle p, H \rangle \subseteq \langle p, \downarrow_s H \rangle$. Here, s denotes the target reduced order. The procedure begins by sorting the columns of H in descending order according to their Euclidean norms. The reduced generator matrix is then partitioned as:*

$$\downarrow_s H = \{H^a \ \ H^b\}, \tag{6.4}$$

where $H^a \in \mathbb{R}^{n\times(s-n)}$ contains the first $s - n$ dominant columns, and $H^b \in \mathbb{R}^{n\times n}$ is a diagonal matrix whose entries are defined as $h_{ii} = \sum_{j=s-n+1}^{m} |H_{ij}|$ for $i = 1, \dots, n$. The term H_{ij} denotes the (i, j)-th element of H.

Lemma 6.2 *Let $\mathcal{Z} = \langle p, H \rangle \subset \mathbb{R}^n$ be an n-dimensional, m-order zonotope. Any point $x = [x_1, \dots, x_n]^T \in \mathcal{Z}$ is enclosed within the minimal axis-aligned bounding box*

$$\text{Box}(\mathcal{Z}) = \{ x : z_i^- \le x_i \le z_i^+, \ i = 1, \dots, n \},$$

whose bounds are computed as:

$$\begin{cases} z_i^- = p_i - \sum_{j=1}^{m} |H_{i,j}|, \\ z_i^+ = p_i + \sum_{j=1}^{m} |H_{i,j}|. \end{cases} \tag{6.5}$$

Here, H_{ij} denotes the (i, j)-th entry of H, and $[p_1, \dots, p_n]^T$ is the center of $\mathcal{Z}$.

6.2.1.2 Problem Formulation

The degradation behavior of lithium-ion batteries is commonly characterized using the dual-polarization equivalent circuit model, expressed as:

$$\begin{cases} V(t) = V_{\text{ocv}}(\text{SOC}(t)) - i(t)R_0 - V_{ap}(t) - V_{cp}(t), \\ V_{ap}(t) = i(t)R_{ap} - \tau_{ap}\dfrac{dV_{ap}(t)}{dt}, \\ V_{cp}(t) = i(t)R_{cp} - \tau_{cp}\dfrac{dV_{cp}(t)}{dt}, \end{cases} \tag{6.6}$$

where $V(t)$ denotes the terminal voltage, and $V_{\text{ocv}}(\text{SOC}(t))$ is the open-circuit voltage (OCV) as a function of the SOC. The terms $V_{ap}(t)$ and $V_{cp}(t)$ represent the Ohmic and concentration polarization voltages, respectively. The applied current is denoted by $i(t)$, R_0 is the internal resistance, and the time constants $\tau_{ap} = R_{ap}C_{ap}$ and $\tau_{cp} = R_{cp}C_{cp}$ correspond to the polarization branches with resistances R_{ap}, R_{cp} and capacitances C_{ap}, C_{cp}.

The SOC evolution is described by the Coulomb counting relation

$$\text{SOC}(t) = \text{SOC}(0) - \frac{\eta}{C_s}\int_0^t i(\tau)\, d\tau, \tag{6.7}$$

where η is the Coulombic efficiency, C_s denotes the available capacity, and SOC(0) is the initial SOC. The dual-polarization circuit representation is illustrated in Figure 6.1.

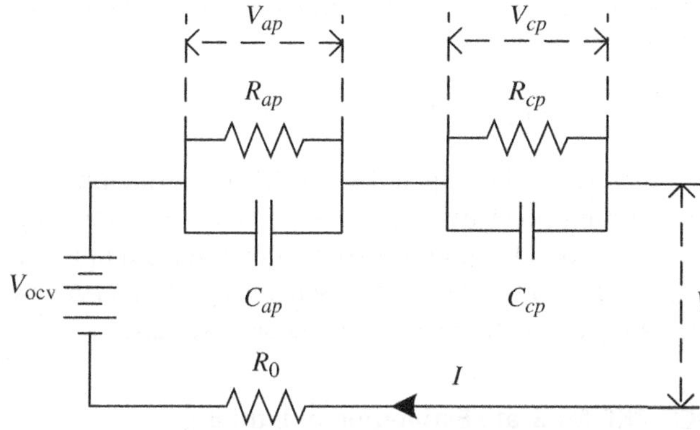

Figure 6.1 Dual-polarization equivalent circuit model of lithium battery degradation.

For constant-current operation, a discrete-time model with sampling period T_s is obtained via zero-order-hold discretization, yielding the state update

$$\begin{bmatrix} \mathrm{SOC}_k \\ V_{ap,k} \\ V_{cp,k} \end{bmatrix} = \begin{bmatrix} 1 & 0 & 0 \\ 0 & e^{-T_s/\tau_{ap}} & 0 \\ 0 & 0 & e^{-T_s/\tau_{cp}} \end{bmatrix} \begin{bmatrix} \mathrm{SOC}_{k-1} \\ V_{ap,k-1} \\ V_{cp,k-1} \end{bmatrix} + \begin{bmatrix} -\dfrac{\eta T_s}{C_s} \\ R_{ap}(1 - e^{-T_s/\tau_{ap}}) \\ R_{cp}(1 - e^{-T_s/\tau_{cp}}) \end{bmatrix} I + w_k, \tag{6.8}$$

where w_k is the process noise.

The terminal voltage at time k is modeled as:

$$V_k = a\,\mathrm{SOC}_k + b - V_{ap,k} - V_{cp,k} - IR_0 + v_k, \tag{6.9}$$

with constants a, b obtained from a linear approximation of the V_{ocv}–SOC curve, and v_k denoting the measurement noise.

The SOH is defined as:

$$\mathrm{SOH}_k = \frac{C_{s,k}}{C_n} \times 100\%, \tag{6.10}$$

where C_n is the nominal-rated capacity and $C_{s,k}$ represents the time-varying available capacity, which monotonically decreases due to degradation. Since it is not directly measurable, we model the capacity evolution as:

$$C_{s,k} = C_{s,k-1} + o_k, \tag{6.11}$$

where o_k denotes the capacity degradation process noise.

Discretizing (6.7) and substituting the SOC relation into the voltage expression, the output model becomes

$$V_k = a\left(\mathrm{SOC}_{k-1} - \frac{\eta I T_s}{C_{s,k}}\right) + b - V_{ap,k} - V_{cp,k} - IR_0 + \upsilon_k. \tag{6.12}$$

The objective of this study is to develop a DCR-ZKF for estimating the degradation-related states of lithium-ion batteries. The approach assumes that the uncertainties w_k, υ_k, and o_k are bounded and represented by the zonotopic sets $w_k \in \mathcal{W} = \langle 0, H_w \rangle$, $\upsilon_k \in \mathcal{V} = \langle 0, H_\upsilon \rangle$, $o_k \in \mathcal{O} = \langle 0, H_o \rangle$, where H_w, H_υ, and H_o are known generator matrices associated with the respective noise processes.

6.2.2 Design of the DCR-ZKF for State Estimation of Lithium Battery Degradation

To jointly estimate SOC and SOH in lithium-ion batteries, this work introduces a zonotopic Kalman filtering framework enhanced with a dynamic complexity-reduction mechanism. Since battery capacity evolves slowly over extended charge–discharge cycles, while SOC varies rapidly at each sampling instant, adopting an identical sampling rate for both SOC and SOH leads to large fluctuations in the estimated capacity and ultimately degrades estimation accuracy. To overcome this limitation, a dual-time-scale structure is employed: a short time scale is used for SOC estimation, whereas SOH is updated on a slower time-scale consistent with its long-term degradation dynamics.

6.2.2.1 SOC Estimation on the Short Time Scale

Let the SOC-related state vector be defined as $x_{c,k} = [\mathrm{SOC}_k, V_{ap,k}, V_{cp,k}]^{\mathrm{T}}$, with input $u_k = I$ and output $y_k = V_k$. Starting from the battery model described in (6.3) and (6.4), the corresponding discrete-time state-space form can be written as:

$$\begin{cases} x_{c,k+1} = A_c x_{c,k} + B_c u_k + w_k, \\ y_k = C_c x_{c,k} + \upsilon_k, \end{cases} \tag{6.13}$$

where $w_k \in \mathcal{W} = \langle 0^{1\times 3}, h_w I_3 \rangle$ and $\upsilon_k \in \mathcal{V} = \langle 0, h_\upsilon \rangle$ denote the process and measurement noises, respectively. The system matrices are

$$A_c = \begin{bmatrix} 1 & 0 & 0 \\ 0 & e^{-T_s/\tau_{ap}} & 0 \\ 0 & 0 & e^{-T_s/\tau_{cp}} \end{bmatrix}, \quad B_c = \begin{bmatrix} -\dfrac{\eta T_s}{C_s} \\ R_{ap}(1 - e^{-T_s/\tau_{ap}}) \\ R_{cp}(1 - e^{-T_s/\tau_{cp}}) \end{bmatrix},$$

$$C_c = \begin{bmatrix} a & -1 & -1 \end{bmatrix}.$$

Given the posterior zonotope estimate at time $k-1$, $\hat{x}_{c,k-1|k-1} \in \langle \hat{p}_{c,k-1|k-1}, \hat{H}_{c,k-1|k-1} \rangle$, the predicted (a priori) zonotope at time k is

$$\hat{x}_{c,k|k-1} \in \hat{\mathcal{X}}_{c,k|k-1} = \langle \hat{p}_{c,k|k-1}, \hat{H}_{c,k|k-1} \rangle,$$

where

$$\hat{p}_{c,k|k-1} = A_c \hat{p}_{c,k-1|k-1} + B_c u_k, \tag{6.14}$$

$$\hat{H}_{c,k|k-1} = \begin{bmatrix} A_c \hat{H}_{c,k-1|k-1} & h_w I_3 \end{bmatrix}. \tag{6.15}$$

At the measurement-update step, incorporation of the voltage measurement refines the estimate to the posterior zonotope

$$\hat{\mathcal{X}}_{c,k|k} = \langle \hat{p}_{c,k|k}, \hat{H}_{c,k|k} \rangle, \tag{6.16}$$

with update equations

$$\hat{p}_{c,k|k} = \hat{p}_{c,k|k-1} + K_{c,k}\left(y_k - C_c \hat{p}_{c,k|k-1}\right), \tag{6.17}$$

$$\hat{H}_{c,k|k} = \begin{bmatrix} (I_3 - K_{c,k} C_c)\hat{H}_{c,k|k-1} & h_v K_{c,k} \end{bmatrix}. \tag{6.18}$$

The zonotopic Kalman gain is determined by

$$K_{c,k} = \frac{\hat{H}_{c,k|k-1} \hat{H}^{\mathrm{T}}_{c,k|k-1} C_c^{\mathrm{T}}}{C_c \hat{H}_{c,k|k-1} \hat{H}^{\mathrm{T}}_{c,k|k-1} C_c^{\mathrm{T}} + h_v^2}. \tag{6.19}$$

To maintain computational tractability, a dynamic complexity-reduction mechanism is introduced. The relative change in the trace of the generator matrix is defined as:

$$\Delta S_k = \frac{\left| \mathrm{tr}\left(\downarrow_s \hat{H}^{\mathrm{T}}_{c,k|k} \downarrow_s \hat{H}_{c,k|k}\right) - \mathrm{tr}\left(\hat{H}^{\mathrm{T}}_{c,k|k} \hat{H}_{c,k|k}\right) \right|}{\mathrm{tr}\left(\hat{H}^{\mathrm{T}}_{c,k|k} \hat{H}_{c,k|k}\right)}, \tag{6.20}$$

where $\downarrow_s \hat{H}_{c,k|k}$ is the reduced-order generator matrix according to Lemma 2, and s denotes the target reduced complexity.

To avoid excessive generator growth across iterations, reduction is triggered whenever the number of generators exceeds a prescribed limit d_m. The updated generator matrix is selected as:

$$\hat{H}^*_{c,k} = \begin{cases} \hat{H}_{c,k|k}, & \text{if } \Delta S_k \geq \varepsilon \text{ or } m > d_m, \\ \downarrow_s \hat{H}_{c,k|k}, & \text{if } 0 < \Delta S_k < \varepsilon, \end{cases} \tag{6.21}$$

where m is the current number of generators and ε is a threshold balancing estimation accuracy and computational effort.

According to Lemma 2, SOC bounds at time k can be extracted from the refined zonotope via interval analysis,

$$\begin{cases} SOC_k^+ = \hat{p}_{c,k|k}(1) + \sum_{j=1}^{m} \left|\hat{H}^*_{c,k}(1, j)\right|, \\ SOC_k^- = \hat{p}_{c,k|k}(1) - \sum_{j=1}^{m} \left|\hat{H}^*_{c,k}(1, j)\right|, \end{cases} \tag{6.22}$$

where m denotes the number of columns in the reduced generator matrix $\hat{H}^*_{c,k}$.

6.2.2.2 SOH Estimation on the Long Time Scale

The nonlinearity introduced by the reciprocal term $\frac{1}{C_{s,k-1}}$ in the voltage model complicates direct estimation of the available capacity. To obtain a tractable linear form, a first-order Taylor expansion around the previous estimate $\hat{C}_{s,k-1}$ is applied, yielding the linearized measurement equation

$$\begin{aligned} V_k &= a\left(\text{SOC}_{k-1} - \frac{\eta I T_s}{C_{s,k-1}}\right) - V_{ap,k} - V_{cp,k} + b - IR_0 + \upsilon_k \\ &\approx -\frac{a\eta I T_s}{\hat{C}_{s,k-1}} + \frac{a\eta I T_s}{\hat{C}^2_{s,k-1}}(C_{s,k-1} - \hat{C}_{s,k-1}) - V_{ap,k} - V_{cp,k} + b - IR_0 + \upsilon_k. \end{aligned} \tag{6.23}$$

Let the long time-scale state be defined as $x_{h,k} = C_{s,k}$, with measurement $y_k = V_k$. The corresponding discrete-time state-space model is

$$\begin{cases} x_{h,k+1} = A_h x_{h,k} + o_k, \\ y_k = C_h x_{h,k} + \upsilon_k, \end{cases} \tag{6.24}$$

where $o_k \in \mathcal{O} = \langle 0, h_o \rangle$ and $\upsilon_k \in \mathcal{V} = \langle 0, h_\upsilon \rangle$. For SOH estimation, the system matrices are $A_h = 1$, $C_h = \frac{a\eta I T_s}{\hat{C}^2_{s,k-1}}$.

Assume that, at time instant $k-1$, the posterior state set is represented by the zonotope $x_{h,k-1|k-1} \in \langle \hat{p}_{h,k-1|k-1}, \hat{H}_{h,k-1|k-1} \rangle$, the prediction and update steps—analogous to those in Section 6.2.2.1—yield the posterior zonotope

$$\hat{\mathcal{X}}_{h,k|k} = \langle \hat{p}_{h,k|k}, \hat{H}_{h,k|k} \rangle, \tag{6.25}$$

where

$$\hat{p}_{h,k|k} = A_h\hat{p}_{h,k-1|k-1} + K_{h,k}\left(y_k - C_hA_h\hat{p}_{h,k-1|k-1}\right),$$
$$\hat{H}_{h,k|k} = \begin{bmatrix} (1 - K_{h,k}C_h)A_h\hat{H}_{h,k-1|k-1} \\ (1 - K_{h,k}C_h)h_o \\ K_{h,k}h_v \end{bmatrix}^{\mathrm{T}}. \tag{6.26}$$

The zonotopic Kalman gain is computed as:

$$K_{h,k} = \frac{\hat{H}_{h,k|k-1}\hat{H}^{\mathrm{T}}_{h,k|k-1}C^{\mathrm{T}}_h}{C_h\hat{H}_{h,k|k-1}\hat{H}^{\mathrm{T}}_{h,k|k-1}C^{\mathrm{T}}_h + h^2_v}. \tag{6.27}$$

The upper and lower bounds of SOH at time k are extracted through interval analysis of the zonotope:

$$\begin{cases} \mathrm{SOH}^+_k = \left(\hat{p}_{h,k|k} + \sum_{i=1}^{m} |\hat{H}_{h,k|k}(i)|\right)/C_n, \\ \mathrm{SOH}^-_k = \left(\hat{p}_{h,k|k} - \sum_{i=1}^{m} |\hat{H}_{h,k|k}(i)|\right)/C_n, \end{cases} \tag{6.28}$$

where m is the number of generators in $\hat{H}_{h,k|k}$.

The complete DCR-ZKF architecture performing joint SOC and SOH estimation is illustrated in Figure 6.2.

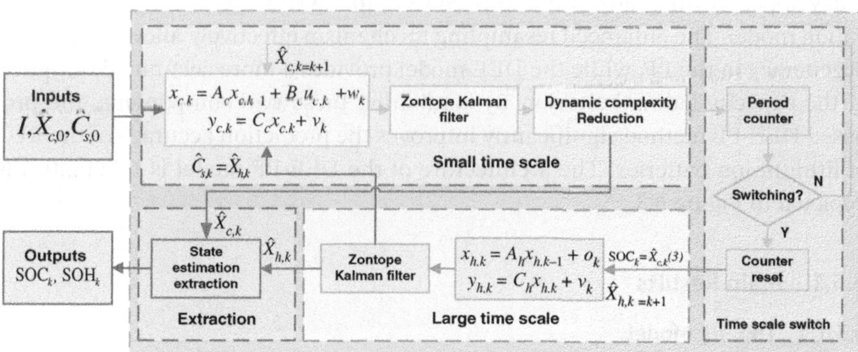

Figure 6.2 Structure of the proposed DCR-ZKF for SOC and SOH estimation in lithium batteries.

6.2.3 Conclusion

In this section, we presented a DCR-ZKF for the joint estimation of SOC and SOH in lithium batteries. The proposed algorithm utilizes a dual time-scale framework, with a short time scale for SOC estimation and a longer one for SOH estimation, effectively decoupling the fast-varying SOC and slow-varying SOH dynamics. This structure minimizes the interference between the two estimations, ensuring that the accuracy of SOC estimation does not degrade the estimation of SOH. The DCR-ZKF also incorporates an adaptive complexity reduction strategy that optimizes the zonotope's shape generator matrix, balancing computational load and estimation accuracy. This allows the algorithm to efficiently process data while maintaining high precision in both state estimations.

The proposed method demonstrates several advantages over traditional zonotopic filtering approaches, particularly in terms of robustness against noise and the accuracy of both SOC and SOH estimates. By dynamically adjusting the complexity of the zonotope representations, the DCR-ZKF reduces computational overhead without compromising the estimation performance. In conclusion, the DCR-ZKF provides an effective solution for the joint estimation of SOC and SOH, making it a promising tool for BMSs, particularly in applications where high accuracy and computational efficiency are required. Further research can explore its integration with other advanced filtering techniques and its application to various industrial systems requiring state estimation under uncertain conditions.

6.3 Double Exponential Empirical Particle Filter

In this section, a double exponential empirical particle filter (DEE-PF) algorithm is developed by integrating an improved resampling strategy with the DEE degradation model. The enhanced resampling mechanism effectively alleviates particle degeneracy in the PF, while the DEE model provides a more accurate description of the capacity-fading behavior. By combining these two components, the proposed DEE-PF method significantly improves the prediction accuracy of the RUL of lithium-ion batteries. The architecture of the DEE-PF model is schematically depicted in Figure 6.3.

6.3.1 Main Results

6.3.1.1 DEE-PF Model

The DEE model has demonstrated strong capability in capturing the nonlinear degradation behavior of lithium-ion battery capacity [91]. Derived from extensive empirical data, the DEE model is expressed as:

$$Q(k) = a(k)\exp\left(b(k)k\right) + c(k)\exp\left(d(k)k\right) + \upsilon(k), \tag{6.29}$$

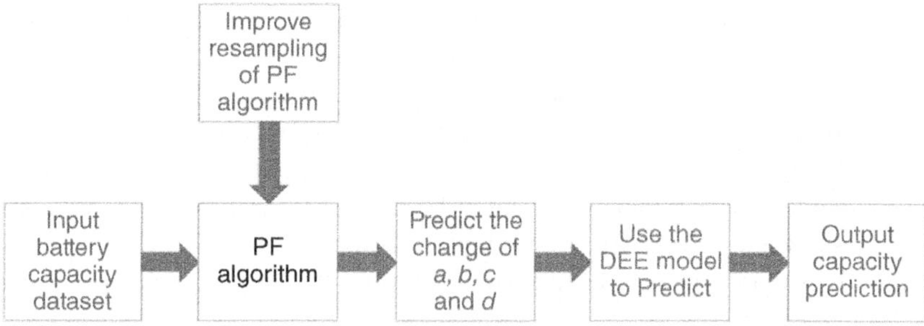

Figure 6.3 Model framework of the proposed DEE-PF algorithm.

Here, the coefficients a and c characterize the internal impedance behavior, while b and d describe the corresponding degradation rates; k denotes the cycle index and $Q(k)$ is the measured capacity. The disturbance term $\upsilon(k)$ is modeled as Gaussian white noise with zero mean and variance σ_υ, i.e., $\upsilon(k) \sim \mathcal{N}(0, \sigma_\upsilon)$.

In the DEE forecasting model, the state vector is chosen as:

$$X(k) = \begin{bmatrix} a(k) & b(k) & c(k) & d(k) \end{bmatrix}, \tag{6.30}$$

and each state component is assumed to evolve according to a random-walk process,

$$\begin{cases} a(k) = a(k-1) + w_a(k-1) \sim \mathcal{N}(0, \sigma_a), \\ b(k) = b(k-1) + w_b(k-1) \sim \mathcal{N}(0, \sigma_b), \\ c(k) = c(k-1) + w_c(k-1) \sim \mathcal{N}(0, \sigma_c), \\ d(k) = d(k-1) + w_d(k-1) \sim \mathcal{N}(0, \sigma_d), \end{cases} \tag{6.31}$$

where $w_a(k-1)$, $w_b(k-1)$, $w_c(k-1)$, and $w_d(k-1)$ denote the corresponding process noise terms, and $\sigma_a, \sigma_b, \sigma_c, \sigma_d$ are their variances, respectively.

From Eq. (6.29), it is evident that once the evolution of parameters a, b, c, and d is accurately inferred, the future capacity trajectory $Q(k)$ can be readily predicted.

In recent studies, particle filtering has become a widely used technique for nonlinear state estimation. PF approximates the posterior density of a stochastic system using a set of discrete weighted samples, replacing integral operations with sample means to obtain minimum-variance state estimates [100]. A standard particle filtering procedure usually comprises the following stages: initialization, importance sampling, weight computation and normalization, resampling, and state estimation. Among these steps, the design of the importance density and the resampling mechanism has a particularly strong influence on filter performance

[101]. Common resampling strategies include multinomial, random, residual, and systematic resampling.

A well-known drawback of the standard PF is particle degeneracy, where after several iterations, only a few particles retain significant weight while most contribute negligibly. This results in loss of particle diversity during resampling, reducing estimation accuracy. Consequently, numerous improved PF variants focus on designing more effective resampling schemes.

A general discrete-time nonlinear system is formulated as follows [94]:

$$x_k = f_k(x_{k-1}, v_{k-1}), \tag{6.32}$$

$$z_k = h_k(x_k, u_k), \tag{6.33}$$

representing the state transition model and observation model, respectively.

The fundamental principle of PF is to approximate the posterior state distribution. The posterior probability density at time k may be approximated by a particle set as:

$$\hat{p}(x_{0:k}|Z_k) = \frac{1}{N}\sum_{i=1}^{N} \delta(x_{0:k} - x_{0:k}^i), \tag{6.34}$$

where $\delta(\cdot)$ is the Dirac delta function. Direct sampling from the true posterior is generally infeasible; hence, an auxiliary distribution $q(x_{0:k}|Z_k)$, known as the importance density, is introduced to facilitate tractable sampling.

The main stages of the improved particle filtering algorithm are summarized as follows.

Step 1: Initialization. Generate the initial particle set $\{x_{0:k}^i \mid Z_k\}$ and assign equal weights $\{w_0^i\}_{i=1,\ldots,N} = \{1/N\}$ according to the known prior distribution $p(x_0)$. Set the cycle index $k = 1$, where N denotes the total number of particles.

Step 2: Importance sampling. Draw samples from the importance density function: $x_k^i \sim q(x_k \mid x_{0:k-1}, Z_k), \qquad i = 1, \ldots, N$.

Step 3: Weight update and normalization. Upon receiving a new observation, update the particle weights as:

$$w_k^i \propto w_{k-1}^i \frac{p(Z_k \mid x_k^i)\, p(x_k^i \mid x_{k-1}^i)}{q(x_k^i \mid x_{0:k-1}^i, Z_k)}, \tag{6.35}$$

followed by normalization:

$$\tilde{w}_k^i = \frac{w_k^i}{\sum_{i=1}^{N} w_k^i}. \tag{6.36}$$

Instead of Eq. (6.35), a Gaussian function is employed to compute particle weights:

$$w(i,k) = \frac{\exp\left(-\left(Z(k) - Z_m(i,k)\right)^2\right)}{2R}, \tag{6.37}$$

where i is the particle index, R is the observation noise covariance, $Z(k)$ is the measured battery capacity at cycle k, $Z_m(i,k)$ is the predicted capacity associated with particle i, and $w(i,k)$ is the resulting weight. The posterior probability density is then approximated by

$$p(x_k \mid Z_k) = \sum_{i=1}^{N} w_k^i \delta(x_k - x_k^i). \tag{6.38}$$

Step 4: Improved resampling. To mitigate particle degeneracy, an enhanced resampling strategy using two thresholds p_1 and p_2 is adopted. Particles with $w(i) > p_1$ are duplicated; those with $w(i) < p_2$ are discarded; particles satisfying $p_2 \le w(i) \le p_1$ are preserved.

The procedure consists of:

1) Sorting the particle set in descending order of weight.
2) Selecting thresholds p_1 and p_2 such that $p_1 > p_2$.
3) Updating the particle set according to

$$X_m(i) = \begin{cases} A, & \text{if } w(i) > p_1, \\ X_m(i), & \text{if } p_2 \le w(i) \le p_1, \\ B, & \text{if } w(i) < p_2, \end{cases} \tag{6.39}$$

where $X_m(i)$ denotes the state of the i-th particle, A denotes duplication, and B denotes elimination.

4) Updating the PF estimate using the average of the M retained particles:

$$X_{pf}(k) = \frac{1}{M} \sum_{i=1}^{M} X_m(i,k), \tag{6.40}$$

where $X_{pf}(k)$ represents the state estimate at cycle k.

Step 5: State estimation. The posterior filtering density can be represented by an empirical distribution of equally weighted particles,

$$\hat{p}(x_k \mid Z_k) = \frac{1}{N} \sum_{i=1}^{N} \delta(x_k - \hat{x}_k^i), \tag{6.41}$$

from which the state estimate is computed as the sample mean

$$\tilde{x}_k = E(x)_{\hat{p}(x_k|Z_k)} = \int x\hat{p}(x_k \mid Z_k)\, dx = \frac{1}{N}\sum_{i=1}^{N} \tilde{x}_k^i. \tag{6.42}$$

Step 6: Iteration. Increase the time index by one, i.e., set $k \leftarrow k + 1$, and repeat the procedure from Step 2. After k reaches N, the estimated trajectories of the parameters a, b, c, and d are obtained. Finally, using Eq. (6.29), the corresponding capacity prediction $Q(k)$ is evaluated.

6.3.1.2 Prediction Steps

In the proposed algorithm, the capacity data of a lithium-ion battery over N charge–discharge cycles are utilized. The first S samples are used for model training, and the remaining $N - S$ samples are employed for prediction and comparison with the ground truth.

Input: Battery capacity measurements over N charge–discharge cycles.

Output: Predicted battery capacities for the last $N - S$ cycles.

Step 1: Use the battery capacity Q as the quantitative index of RUL, and scale the initial capacity such that the normalized starting value equals 1.

Step 2: Fit the measured capacity-fade trajectory with the DEE model and, from this regression, obtain the initial parameter estimates a_0, b_0, c_0, and d_0 for the parameter vector (a, b, c, d).

Step 3: Set up the PF framework by defining the particle set X_m, the filter state X_{pf}, the capacity measurement sequence Z, the particle-level capacity predictions Z_m, the filter-level predicted capacity Z_{pf}, and the particle weights W. The set X_m contains M particles indexed by $m = 1, 2, \dots, M$. The PF recursion is performed until the iteration counter k reaches N. The initial particles are generated by perturbing the nominal parameters a_0, b_0, c_0, and d_0 with process noise.

Step 4: Initialize the PF cycle index with $k = 2$. At cycle k, draw or propagate the M particles, substitute each particle state into the DEE model to obtain the corresponding capacity prediction, compute the importance weight of every particle according to the prediction error, and finally normalize the weight set W.

Step 5: Apply the improved resampling scheme with thresholds p_1 and p_2 (with $p_1 > p_2$): duplicate particles whose weights exceed p_1, discard particles whose weights are below p_2, and retain the remaining particles. After replication, elimination, and retention, update the PF state X_{pf}, and use this updated state to compute the filtered capacity prediction Z_{pf}.

Step 6: Increase the cycle counter by one, i.e., set $k \leftarrow k + 1$, and iterate Steps 4–5 until $k = N$. After this recursion is completed, the PF state X_{pf} and the corresponding filtered capacity sequence Z_{pf} over all observed cycles are obtained.

Step 7: Define the starting index of the prediction horizon as $S = N - \text{FC}$, where FC is the number of cycles to be forecast and N is the total number of measured

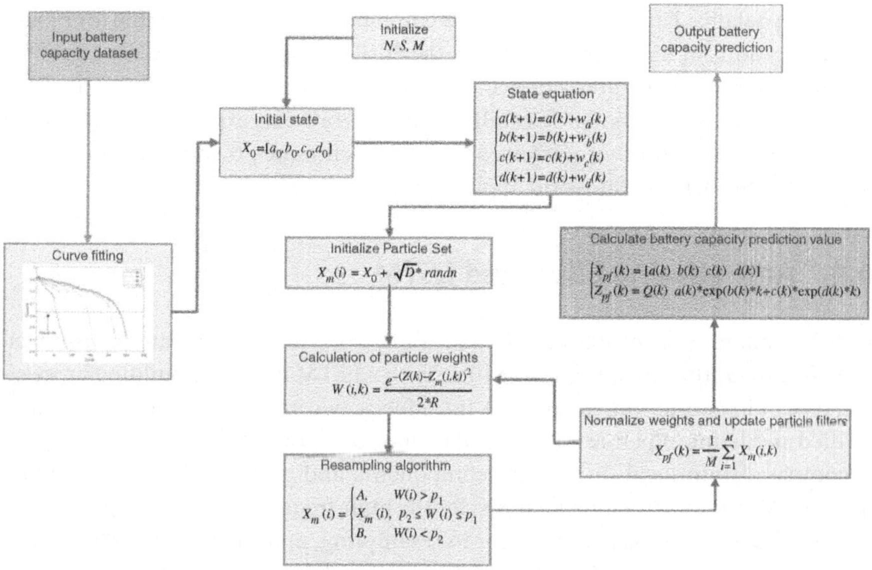

Figure 6.4 Algorithm flowchart of the proposed DEE-PF prediction procedure.

cycles. Using the final PF state X_{pf} from Step 6, identify the DEE model parameters a, b, c, and d. The DEE-PF model is then run forward from cycle S over the prediction horizon to generate the capacity predictions Z_{pf}, from which the RUL of the lithium-ion battery is evaluated.

The overall algorithm flowchart is depicted in Figure 6.4.

6.3.2 Conclusion

In this section, we introduced the DEE-PF algorithm, a robust method for predicting the RUL of lithium-ion batteries. The DEE-PF algorithm combines the strengths of the PF approach with a DEE degradation model, addressing key challenges in predicting battery life. The key contribution of the DEE-PF method lies in the novel resampling strategy, which mitigates particle degradation and enhances particle diversity, leading to more accurate state estimation. By incorporating the DEE model, which simplifies the computation of battery degradation parameters, the DEE-PF algorithm maintains high predictive accuracy while requiring fewer parameters compared to other methods.

The method is particularly advantageous in dealing with non-Gaussian and nonlinear systems, where traditional algorithms often struggle. Moreover, it achieves better prediction performance by maintaining particle diversity and improving the resampling process. The proposed algorithm is computationally efficient, offering a balance between accuracy and complexity, making it suitable for practical applications in BMSs. In conclusion, the DEE-PF algorithm presents a promising approach for RUL prediction in lithium-ion batteries, providing both high accuracy and computational feasibility. Future work can explore the integration of this method with other advanced filtering techniques to further enhance its robustness in various real-world applications.

6.4 Anti-aliasing Filter and LSTM

In this section, a novel RUL prediction framework for power batteries is developed by integrating an anti-aliasing filter with a LSTM network. Building on existing research, the proposed method introduces two main innovations within a unified modeling pipeline: (i) the SOH time series of the power battery is first decomposed into local fluctuation components and a global degradation trend using EEMD and correlation analysis; for the high-frequency components (HFCs), an anti-aliasing filter is applied prior to resampling, and the resampled sequence is then predicted by an LSTM model, which reduces the effective fluctuation frequency and significantly alleviates the low prediction accuracy traditionally associated with HFCs; (ii) before the resampling step, a Monte Carlo-based preprocessing procedure is employed on the HFC data to suppress the influence of outliers, thereby further enhancing the prediction accuracy of this component and improving the overall robustness of the RUL prediction framework.

6.4.1 Problem Formulation

In this section, the SOH sequence is first decomposed into a global degradation trend and multiple fluctuation components using EEMD combined with correlation analysis [102]. The low-frequency component is directly predicted using an LSTM network, whereas the HFCs require additional processing to improve prediction accuracy. Specifically, the HFCs are first preprocessed to constrain outliers within a reasonable range. The preprocessed sequence is then resampled using an anti-aliasing filter to reduce its fluctuation frequency before being predicted by the LSTM. After LSTM prediction, inverse resampling and inverse preprocessing are applied to recover the original frequency characteristics. Finally, the predicted SOH is obtained by integrating the reconstructed HFCs with the predicted main degradation trend. The overall RUL prediction framework based on the anti-aliasing filter and LSTM is shown in Figure 6.5.

6.4.1.1 HFC Prediction Model

HFCs typically contain outliers, and direct resampling may amplify their adverse influence on prediction performance. Therefore, it is necessary to preprocess the HFCs prior to the resampling operation.

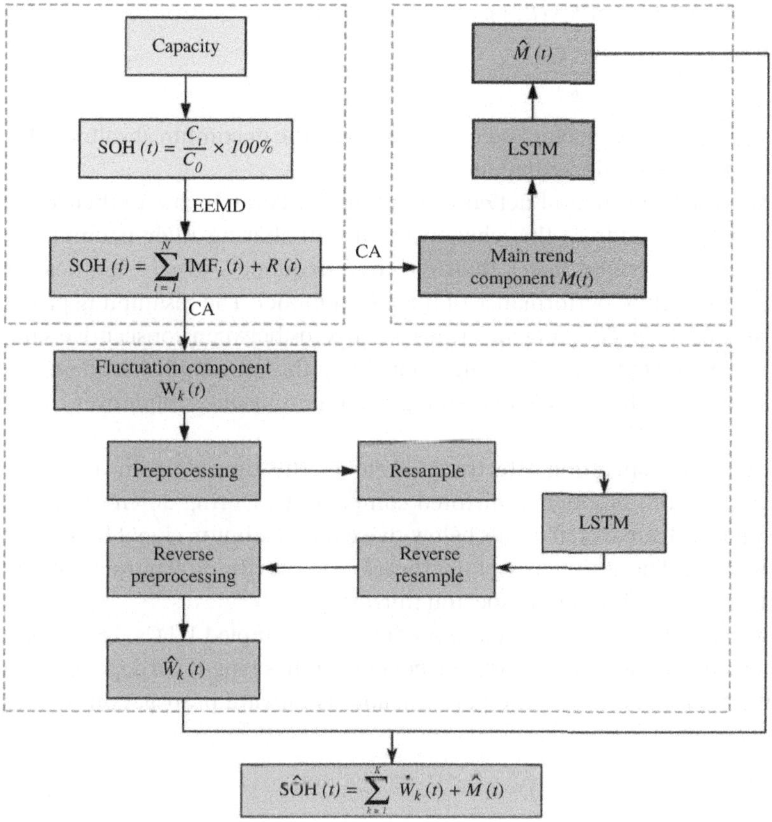

Figure 6.5 Prediction framework of power battery RUL based on anti-aliasing filtering and LSTM.

To begin, an interval threshold is initialized for each fluctuation component as threshold(k), $k = 1, 2, \ldots, K$, where K denotes the number of fluctuation components. Then, the corresponding interval boundaries boundary are computed to satisfy

$$\text{boundary}(k, 1) = -\text{boundary}(k, 2), \tag{6.43}$$

$$p(k) = \text{threshold}(k), \tag{6.44}$$

where boundary(k, 1) and boundary(k, 2) denote the lower and upper bounds of the k-th fluctuation sequence, respectively, and $p(k)$ is the proportion of points lying within the interval. Outliers are then identified and processed according to

$$W_k(\text{error}_k) \notin [\text{boundary}(k, 1), \text{boundary}(k, 2)], \tag{6.45}$$

$$W'_k(\text{error}_k) = \frac{W_k(\text{error}_k)}{\max(k)} \times \text{boundary}(k, 2), \tag{6.46}$$

where error_k is the outlier vector and $\max(k)$ denotes the maximum absolute value of W_k in the k-th fluctuation component.

HFCs are difficult for neural networks to predict accurately due to their rapid variability [103]. To address this challenge, an anti-aliasing filter is employed to resample the HFCs and lower their effective frequency, which significantly improves the prediction performance of the LSTM model. The resampling procedure consists of: (i) *Up-sampling*: inserting $p-1$ zeros between consecutive samples of the original signal; (ii) *Filtering*: convolving the up-sampled sequence y_1 with an anti-aliasing filter h; (iii) *Down-sampling*: retaining one sample every $q-1$ points.

The up-sampling operation effectively stretches the time domain and compresses the spectrum, producing mirrored components. During down-sampling, the sampling rate decreases; if it falls below twice the maximum signal frequency, aliasing will occur. The anti-aliasing filter therefore serves the dual purpose of suppressing aliasing and eliminating spectral mirroring.

An LSTM network is then applied to predict the resampled HFCs. LSTM, as a specialized recurrent neural network, effectively handles long-term dependencies [95, 96]. Its structure is shown in Figure 6.6, and its forward propagation is given by Eqs. (6.47)–(6.52).

$$f_t = \sigma(W_f[h_{t-1}, x_t] + b_f), \tag{6.47}$$

$$i_t = \sigma(W_i[h_{t-1}, x_t] + b_i), \tag{6.48}$$

$$\widetilde{C}_t = \tanh(W_c[h_{t-1}, x_t] + b_c), \tag{6.49}$$

$$C_t = f_t \odot C_{t-1} + i_t \odot \widetilde{C}_t, \tag{6.50}$$

$$h_t = o_t \odot \tanh(C_t), \tag{6.51}$$

$$o_t = \sigma(W_o[h_{t-1}, x_t] + b_o), \tag{6.52}$$

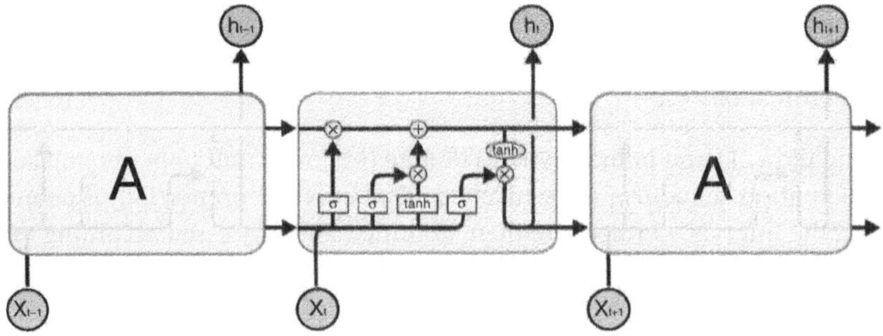

Figure 6.6 Structure of the LSTM network.

where h_t and C_t denote the short-term and long-term memory states, $\widetilde{C}_t$ is the candidate cell state, and x_t is the current input vector. The matrices W_f, W_i, and W_o are associated with the forget, input, and output gates, respectively, and b_f, b_i, b_o are the corresponding bias vectors.

Since the LSTM predictions do not match the original frequency characteristics and processed outliers must be restored, inverse resampling and inverse preprocessing are necessary to obtain the final predicted HFCs. Inverse resampling swaps the stretching factor p and compression factor q, restoring the original frequency scale. Inverse preprocessing restores outlier amplitudes via

$$\hat{W}_k(\text{error}_k) = \frac{\hat{W}_k(\text{error}_k)}{\text{boundary}(k, 2)} \times \max(k). \tag{6.53}$$

6.4.1.2 Monte Carlo-based Optimization of Interval Thresholds

The interval thresholds used in the outlier preprocessing stage can, in principle, be selected empirically. However, the prediction of the fluctuation components is highly sensitive to these thresholds, and experience-based tuning introduces substantial uncertainty, which can significantly degrade prediction accuracy. To address this issue, a Monte Carlo strategy is adopted to optimize the interval thresholds:

$$TH_i = \text{random}\left[N(\mu, \sigma^2)\right], \tag{6.54}$$

$$[TH_{o,1}, \dots, TH_{o,n}] = \arg\min f(TH_1, \dots, TH_n), \tag{6.55}$$

$$\text{RMSE} = \sqrt{\frac{1}{N}\sum_{i=1}^{N} (y_i - \hat{y}_i)^2}, \tag{6.56}$$

where TH_i denotes the candidate interval threshold, $f(\cdot)$ is the root-mean-error (RMSE) of the Monte Carlo-based resampled LSTM (MC-R-LSTM) model corresponding to a given set of thresholds, and $TH_{o,i}$ represents the optimal interval threshold.

Since empirically chosen thresholds are neither systematic nor guaranteed to be optimal, they may significantly compromise the prediction performance. In contrast, determining the optimal thresholds via Monte Carlo search allows the model to more accurately capture the fluctuation characteristics, thereby further improving the prediction accuracy of the RUL of power batteries.

6.4.1.3 SOH and RUL Prediction

The final SOH prediction at time t is obtained by combining the predicted main trend (MT) and the reconstructed fluctuation components:

$$\hat{\mathrm{SOH}}(t) = \sum_{k=1}^{K} \widehat{W}_k(t) + \widehat{M}(t), \tag{6.57}$$

where $\hat{M}(t)$ is the predicted MT component, $\hat{W}_k(t)$ is the predicted fluctuation component of the k-th group, and K denotes the total number of fluctuation components.

When the SOH decays to a predefined failure threshold, the power battery is considered to have reached its EOL. The RUL is defined as the time interval between the current SOH and the failure threshold. Specifically, RUL is given by where T_1 is the cycle index at which the battery hits the failure threshold, and T_2 is the current cycle index.

6.4.2 Conclusion

In this section, we introduced a novel prediction model for the RUL of power batteries, integrating an anti-aliasing filter with LSTM networks. The model addresses the challenges of HFCs in the battery's SOH time series, which often lead to poor prediction accuracy when handled by traditional neural network methods. By employing ensemble empirical mode decomposition (EEMD) and correlation analysis, we effectively decomposed the SOH time series into MT and fluctuation components. The low-frequency MT component was predicted directly using LSTM, while the high-frequency fluctuation component was preprocessed using the Monte Carlo method and resampled through an anti-aliasing filter. This dual approach allowed for improved prediction accuracy by reducing the fluctuation frequency and mitigating the impact of outliers. The final prediction of the SOH and RUL was obtained by integrating the predicted results of both the MT and fluctuation components. This method not only enhances the prediction accuracy of the power battery RUL but also provides a robust framework for dealing with nonlinear and nonstationary data typical of battery degradation.

In conclusion, the proposed anti-aliasing filter and LSTM-based prediction model significantly improve the accuracy and reliability of RUL predictions for power batteries. The method's ability to handle complex, high-frequency fluctuations in the SOH time series makes it an effective tool for BMSs. Future work may focus on extending this approach to other areas of predictive maintenance, such as wind speed forecasting and power plant output prediction, where similar challenges exist.

6.5 Case Study on Battery Remaining Useful Life Analysis

6.5.1 Case 1: DCR-ZKF

To demonstrate the performance and validity of the proposed DCR-ZKF scheme, simulation experiments were conducted using the dual-polarization equivalent circuit model. The associated model parameters are summarized in Table 6.1. The sampling interval is set to $T_s = 5$ s, and the linearized relationship between the OCV and SOC is approximated as $V_{\text{ocv}}(\text{SOC}(t)) = 7.17\,\text{SOC}(t) + 4.12$.

A representative discharge interval is selected to evaluate the SOC estimation performance. To further illustrate the superiority of the proposed method, comparisons are also carried out with a traditional complexity-reduced zonotopic filter and a standard ellipsoidal filter. The initial state, input, and uncertainty descriptions are specified as $x_{c,0} = [1, 5, 5]^{\text{T}}$, $u_k = 1$, $\hat{x}_{c,0} \in \hat{\mathcal{X}}_{c,0} = \langle \hat{p}_{c,0}, \hat{H}_{c,0} \rangle = \langle [1, 5, 5]^{\text{T}}, 10^{-1} I_3 \rangle$, $w_k \in \langle 0, H_w \rangle = \langle [0, 0, 0]^{\text{T}}, 10^{-3} I_3 \rangle$, and $v_k \in \langle 0, H_v \rangle = \langle 0, 10^{-4} \rangle$.

The system matrices for SOC estimation are

$$A_c = \begin{bmatrix} 1 & 0 & 0 \\ 0 & 0.8969 & 0 \\ 0 & 0 & 0.0021 \end{bmatrix}, \quad B_c = \begin{bmatrix} -0.0244 \\ 0.0031 \\ 0.0399 \end{bmatrix}, \quad C_c = \begin{bmatrix} 7.17 & -1 & -1 \end{bmatrix}.$$

Figure 6.7 compares the SOC estimates obtained by the proposed DCR-ZKF, the traditional zonotopic filter, and the ellipsoid filter. The true SOC trajectory is plotted as a solid line, while the dashed, dotted, and dash-dotted curves correspond to the DCR-ZKF, zonotopic filter, and ellipsoid filter, respectively. As shown in Figure 6.7, the DCR-ZKF consistently achieves the highest accuracy, demonstrating superior tracking capability.

Figure 6.8 compares the estimation bounds at time steps $k = 50$, 100, and 200. The true state is marked by a star. As illustrated in Figure 6.8, the DCR-ZKF provides significantly tighter bounds than both the traditional zonotopic filter and the ellipsoidal filter, highlighting the benefit of DCR.

Table 6.1 Equivalent circuit parameters.

Symbol	Physical meaning	Value
R_0	Ohmic internal resistance	0.52 Ω
R_{ap}	Activation polarization resistance	0.03 Ω
R_{cp}	Concentration polarization resistance	0.04 Ω
C_{ap}	Activation polarization capacitance	1531.2 F
C_{cp}	Concentration polarization capacitance	20.21 F
C_n	Nominal capacity	205 A · h

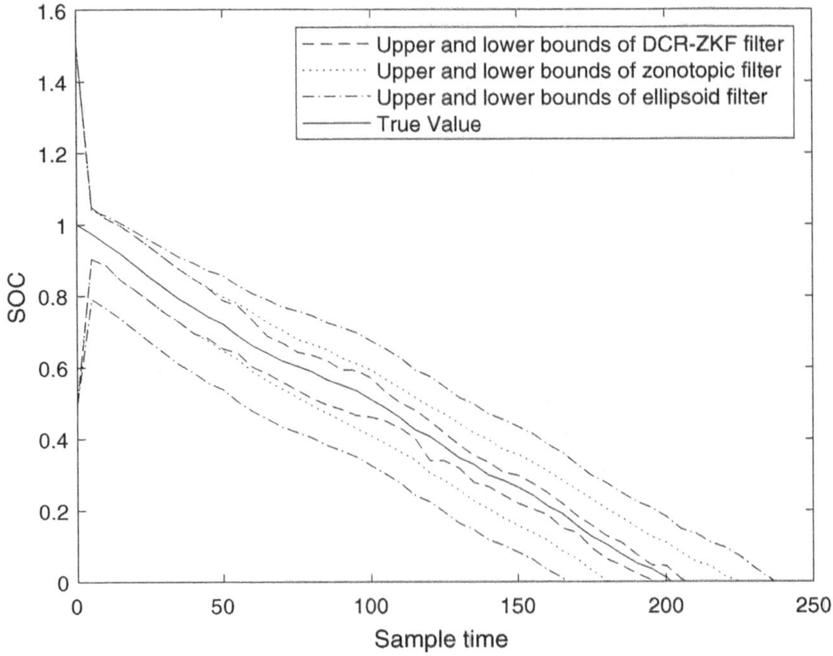

Figure 6.7 Comparison of the bounds of SOC obtained using different filters.

For SOH estimation, the initial conditions are chosen as $x_{h,0} = 102.5$ and $\hat{x}_{h,0} \in \hat{\mathcal{X}}_{h,0} = \langle 102.5, 4.1 \rangle$. The zonotopic uncertainty bounds are set as $o_k \in \langle 0, 10^{-3} \rangle$ and $v_k \in \langle 0, 10^{-4} \rangle$, and the system parameters are given by $A_h = 1$ and $C_h = 35.85/\hat{p}^2_{s,k-1|k-1}$.

Figure 6.9 presents the SOH estimation results from the three filters. Although the performance improvement of DCR-ZKF is less pronounced for SOH than for SOC—owing to the inherently slow-varying nature of capacity degradation—it still outperforms the ellipsoid filter and provides slightly tighter bounds than the conventional zonotopic filter. The improved SOC accuracy achieved by DCR-ZKF indirectly enhances SOH estimation quality, confirming the benefit of the integrated dual-time-scale design.

6.5.2 Case 2: DEE-PF

The experimental validation in this study was carried out using lithium-ion battery capacity degradation data measured under laboratory conditions at the University of Maryland. The dataset contains four groups of charge–discharge records, labeled A3, A5, A8, and A12. The corresponding capacity fading trajectories are shown in Figure 6.10, where the end of life is defined as the point at which

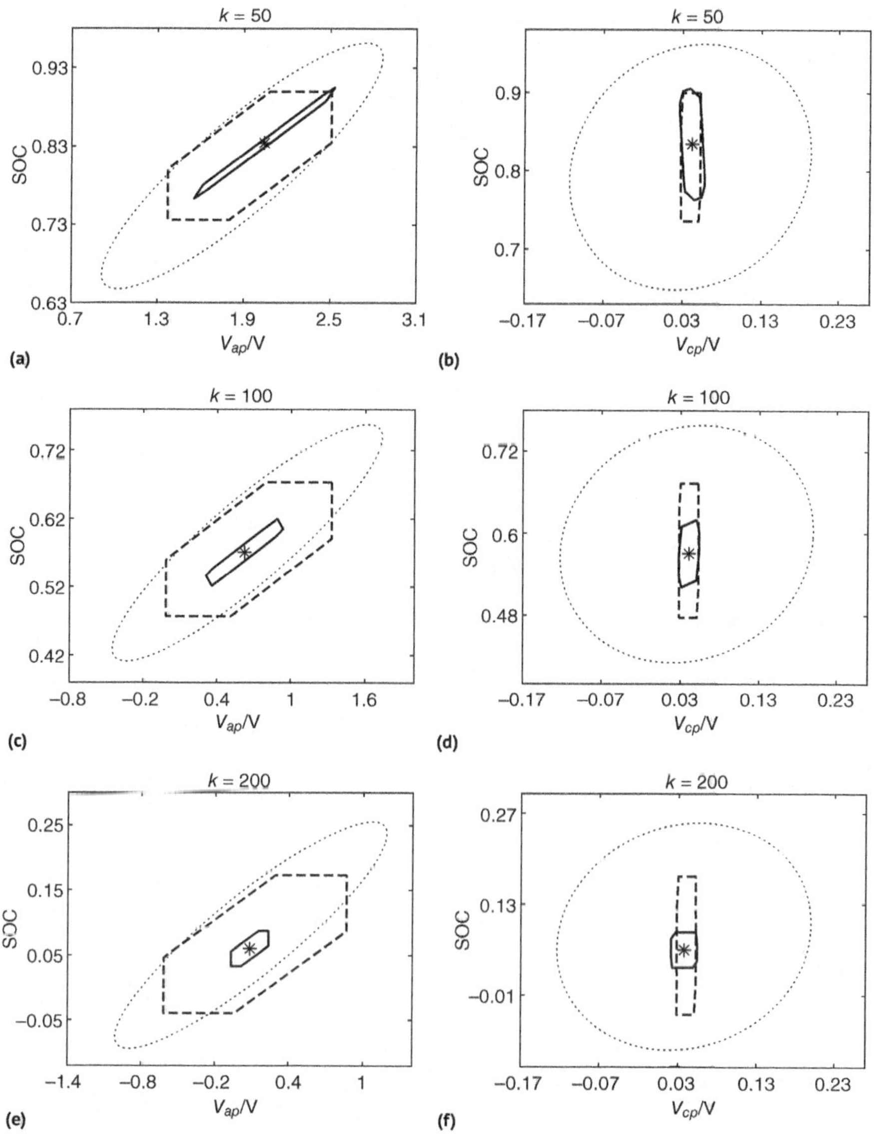

Figure 6.8 Comparison of actual states and estimated states by different filters (dotted boundary: ellipsoid filter; dashed boundary: zonotopic filter; solid boundary: DCR-ZKF; star: true state).

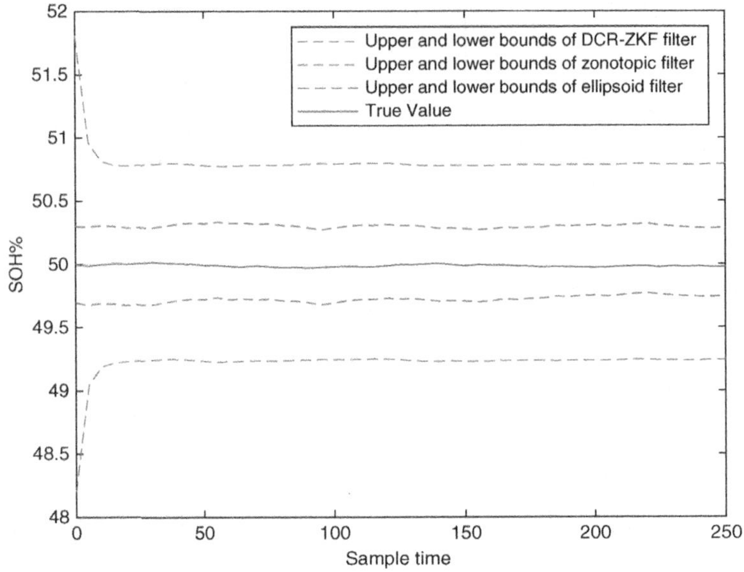

Figure 6.9 Comparison of the bounds of SOH obtained using different filters.

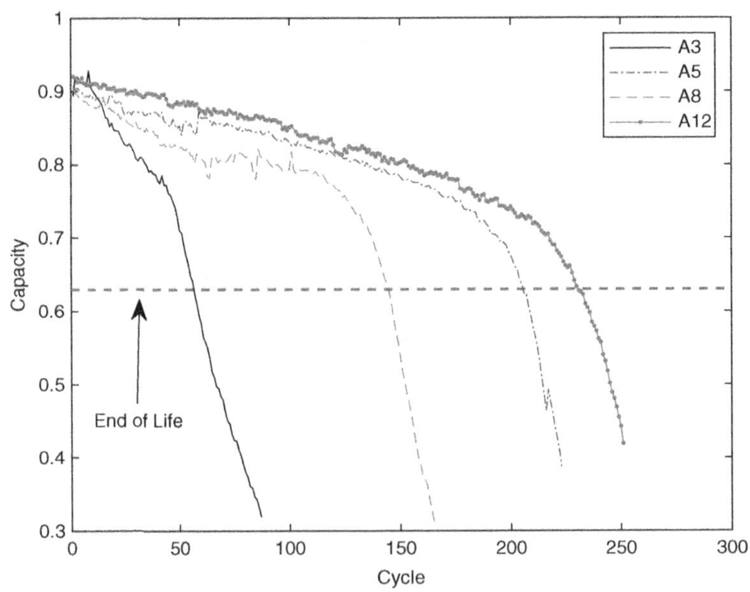

Figure 6.10 Battery capacity attenuation for four cells (A3, A5, A8, and A12).

the capacity drops to 70% of its initial value. The initial DEE model parameters $a_0 = -0.0000083499$, $b_0 = 0.055237$, $c_0 = 0.90097$, and $d_0 = -0.00088543$ are obtained by fitting three sets of experimental data.

For a detailed assessment, the capacity degradation over 200 cycles of cell A12 is used as the sample sequence. The initial 180 cycles are employed to train the model, whereas the subsequent 20 cycles are reserved for prediction and validation. After performing parameter identification with the PF-based scheme, the time-varying trajectories of the DEE parameters a, b, c, and d are obtained, as illustrated in Figure 6.11. The resulting DEE-PF capacity prediction for A12 is illustrated in Figure 6.12, where the predicted trajectory closely follows the measured capacity, indicating small prediction errors and satisfactory tracking performance.

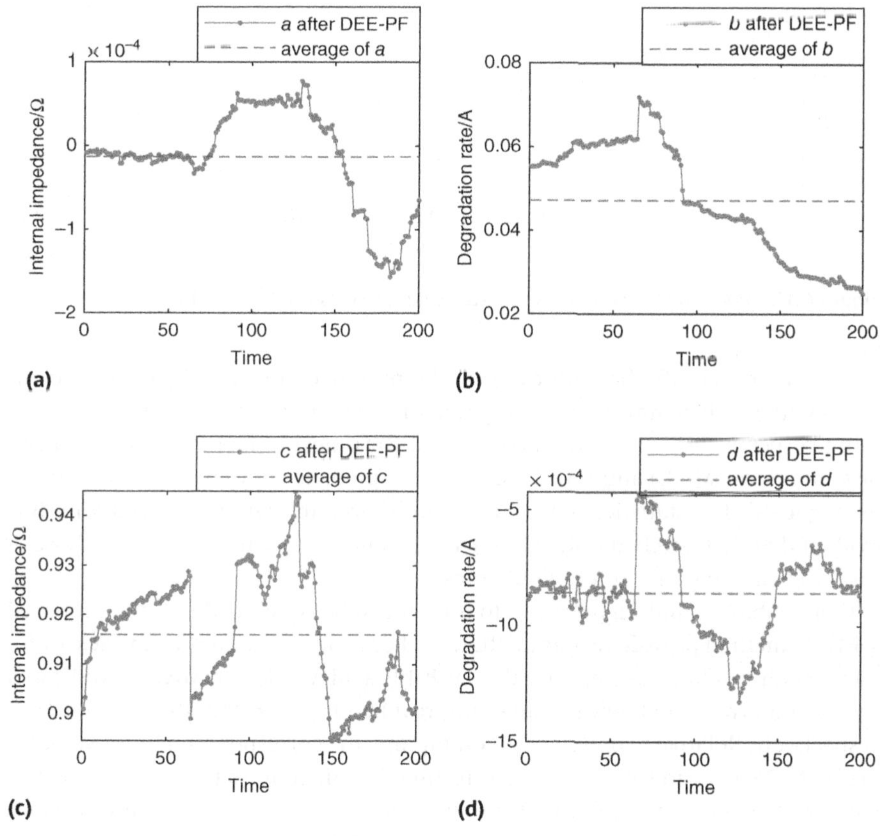

Figure 6.11 Temporal evolution of DEE model parameters *a*, *b*, *c*, and *d*.

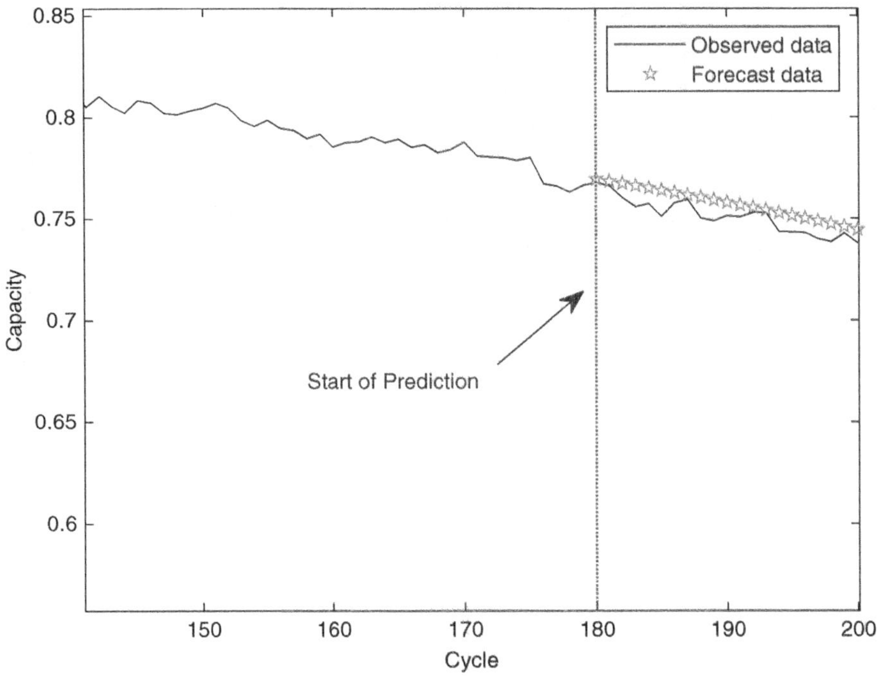

Figure 6.12 Prediction results of A12 battery capacity using the DEE-PF method.

To further quantify the performance, the proposed DEE-PF algorithm is compared with two benchmark PFs using random resampling and polynomial resampling, respectively. Each method is executed 100 times, and for each run the mean absolute error (MAE) and RMSE between the predicted and measured capacities are recorded. The statistical results are summarized in Figures 6.13 and 6.14. It is evident that both MAE and RMSE of DEE-PF are consistently lower than those of the two benchmark resampling schemes.

From Table 6.2 and Figures 6.12 to 6.14, it can be concluded that the proposed DEE-PF method provides accurate RUL prediction for lithium-ion batteries. Compared with random resampling, the DEE-PF achieves lower MAE, RMSE, and maximum error (ME), with relative improvements of 19.79%, 18.34%, and 60%, respectively. Relative to polynomial resampling, the corresponding improvements are 25%, 25.68%, and 77.24%. These results demonstrate that, in the context of capacity fade prediction, the DEE-PF-based approach can substantially enhance prediction accuracy compared with conventional PF resampling schemes.

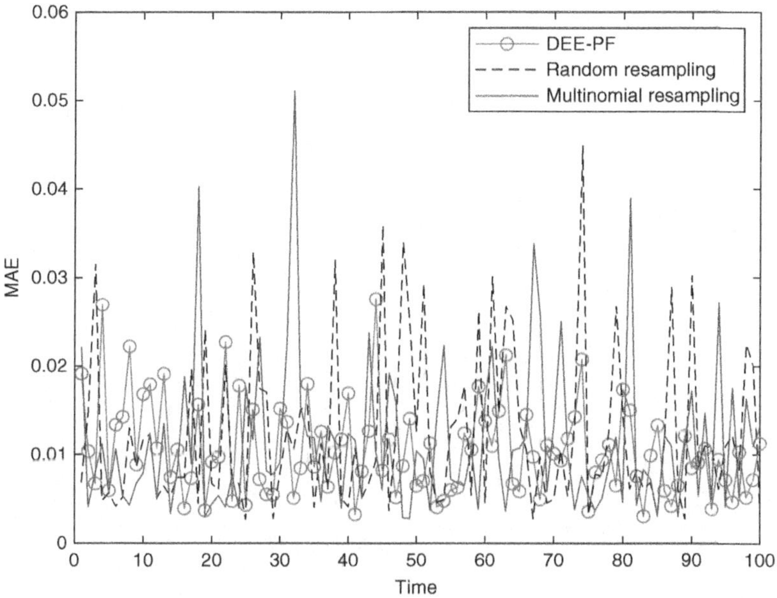

Figure 6.13 MAE comparison.

6.5.3 Case 3 MC-R-LSTM

To evaluate the performance of the algorithm developed in Section 6.4, simulation experiments are carried out using the NASA lithium-ion battery dataset. The dataset consists of several groups of measurements from cells with a rated capacity of 2 A · h, including charge, discharge, and impedance curves collected at room temperature.

Here, Battery #5 is selected to illustrate the complete algorithmic workflow and prediction performance. The discharge-capacity sequence of Battery #5 is extracted as the SOH degradation series, from which the SOH values are computed. The proposed MC-R-LSTM method is evaluated and compared against three benchmark methods: LSTM, MT, and Resample-LSTM (R-LSTM).

6.5.3.1 SOH Decomposition

The SOH sequence is first decomposed into intrinsic mode functions (IMFs) and a residual trend component using EEMD. The decomposition results are presented in Figure 6.15.

After EEMD processing, the SOH is decomposed into six IMFs and one residual component. IMFs 1–4 exhibit distinct oscillatory behavior, whereas IMFs 5–6 show weaker fluctuations, and the residual component captures the long-term

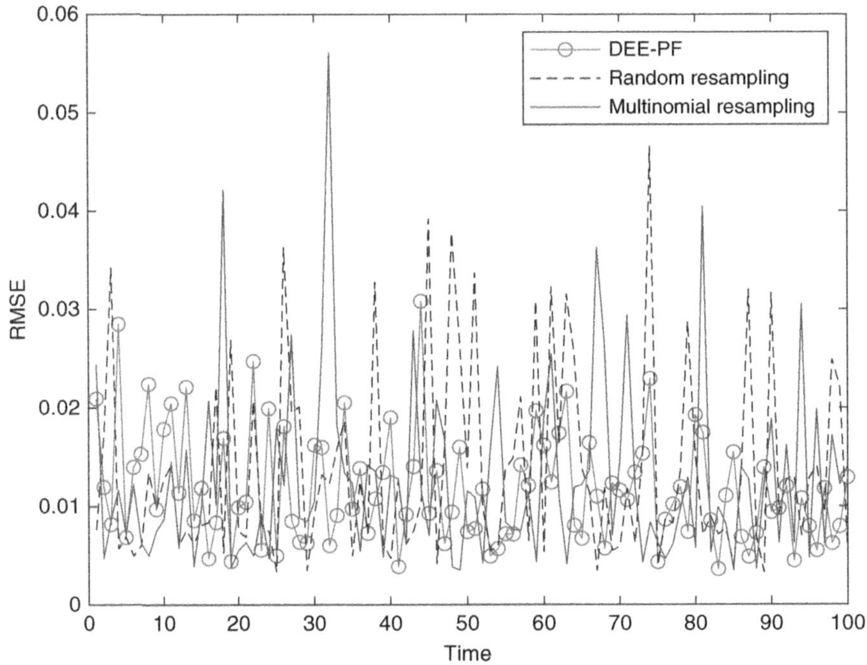

Figure 6.14 RMSE comparison.

Table 6.2 Comparison of prediction errors.

Metrics	DEE-PF	Random	Polynomial
MAE	0.0096	0.0115	0.0122
RMSE	0.0109	0.0129	0.0137
ME	0.0435	0.0696	0.0771

degradation trend. Pearson correlation analysis is then conducted between each IMF and the original SOH sequence, with the results summarized in Table 6.3.

Given the correlation threshold of 0.3931, IMFs 1–4 are identified as fluctuation components, while IMFs 5–6, together with the residual form the MT component. This decomposition ensures the effective separation of global degradation and local oscillations, mitigating the adverse effect of mode mixing on the prediction of the underlying SOH degradation trend.

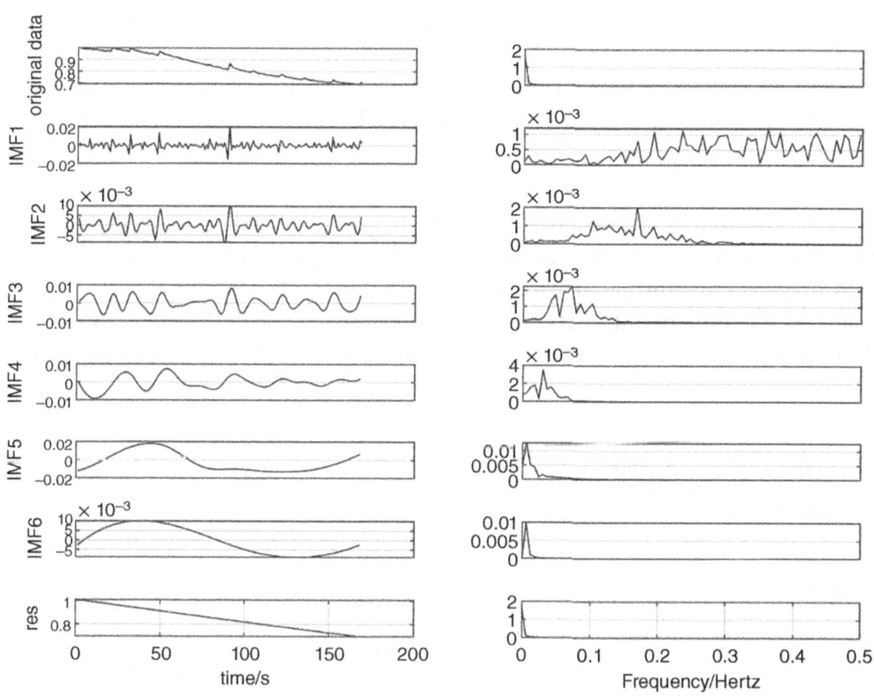

Figure 6.15 EEMD decomposition of the SOH sequence.

Table 6.3 Pearson correlation coefficients between IMFs and SOH.

IMFs	**1**	**2**	**3**	**4**	**5**	**6**
r	0.0150	0.0622	0.1246	−0.1702	0.5730	0.8709

Table 6.4 Parameters of the LSTM network.

Input	**Output**	**Cell**	**Rate**	**Iteration**
1	1	20	0.01	500

6.5.3.2 Main Trend Prediction

The LSTM network used to predict the MT component is configured according to the parameters listed in Table 6.4. Specifically, the first 60% of the MT sequence is used as the training set, and the remaining 40% is reserved as the test set. The LSTM model is trained on the training set with the settings in Table 6.4, and its prediction performance is subsequently evaluated on the test set.

The prediction results are shown in Figure 6.16. The MT exhibits a smooth downward trajectory, and the LSTM-based prediction closely follows this behavior. This indicates that the LSTM model achieves a high-quality fit to the underlying degradation trend and that its forecasts are largely consistent with the actual long-term SOH evolution.

6.5.3.3 Fluctuation Component Prediction

The mean u and standard deviation σ used for generating candidate interval thresholds via the normal distribution are listed in Table 6.5.

For each candidate interval threshold set generated by the Monte Carlo procedure, the corresponding RMSE is computed, and the set yielding the minimum

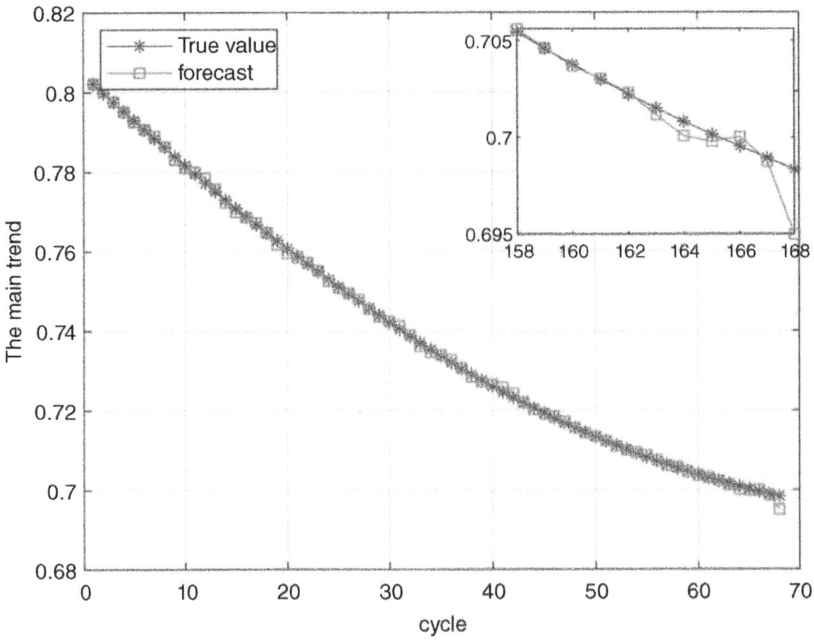

Figure 6.16 Prediction results of the MT component.

Table 6.5 Parameters of the normal distribution.

Number of HFCs	1	2	3	4
u	0.4	0.6	0.8	0.99
σ	0.05	0.05	0.05	0.05

Table 6.6 Interval threshold sets.

Number of HFCs	1	2	3	4
R-LSTM	0.5	0.6	0.7	0.8
MC-R-LSTM	0.22	0.43	0.60	0.98

RMSE is selected as the optimal threshold configuration. Table 6.6 compares the interval thresholds obtained by the proposed MC-R-LSTM method with those used in the baseline R-LSTM model.

After outlier preprocessing, each fluctuation component is compressed into the interval defined by its optimized threshold. During resampling, the time-domain amplification factor is set to $p = 5$ and the time-domain compression factor to $q = 1$. Consequently, the original sequence is stretched by a factor of five in time, reducing its effective frequency by the same factor. An anti-aliasing filter is then applied to obtain a resampled, lower-frequency sequence. In the reverse resampling stage, the factors are switched to $p = 1$ and $q = 5$, restoring the original frequency characteristics by compressing the prediction in time by a factor of five.

The resampled fluctuation components are predicted using an LSTM network, with the first 60% of the samples used for training and the remaining 40% for testing. After prediction, the outlier values are restored via inverse preprocessing, and the recovered sequence represents the final predicted fluctuation component. The prediction results and errors under different models are shown in Figures 6.17 and 6.18.

As shown in Figure 6.18, the prediction errors of both R-LSTM and MC-R-LSTM predominantly remain within the interval $[-0.005, 0.005]$. Notably, the MC-R-LSTM model produces fewer deviations beyond this range, whereas the LSTM method exhibits more substantial errors, with some values even exceeding the interval $[-0.01, 0.01]$. These results underscore the superiority of the proposed MC-R-LSTM method in accurately predicting high-frequency fluctuation components.

6.5.3.4 Prediction of SOH and RUL

By combining the predicted MT and the reconstructed fluctuation components, the final SOH prediction is obtained. The corresponding results are presented in Figures 6.19 and 6.20.

As shown in Figure 6.19, the MT and LSTM methods exhibit noticeable deviations at several time points, yielding significantly larger overall errors compared with R-LSTM and MC-R-LSTM. Between the two resampling-based approaches, MC-R-LSTM achieves the smallest overall prediction error, demonstrating superior accuracy.

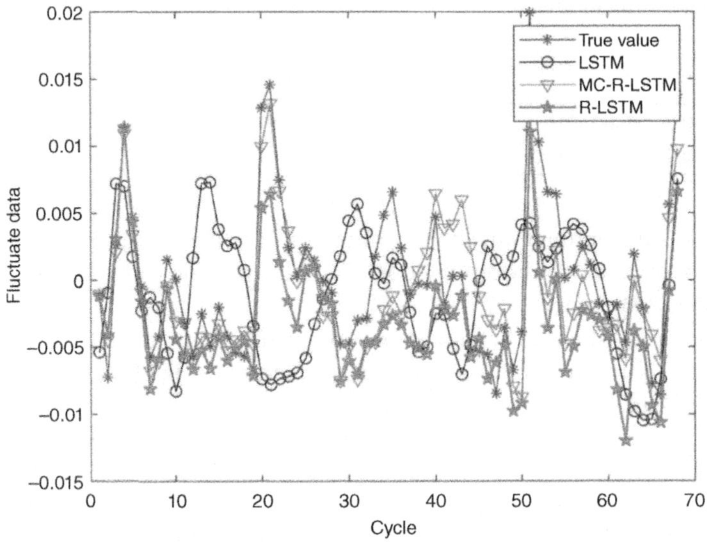

Figure 6.17 Prediction comparison of the fluctuation component under different models.

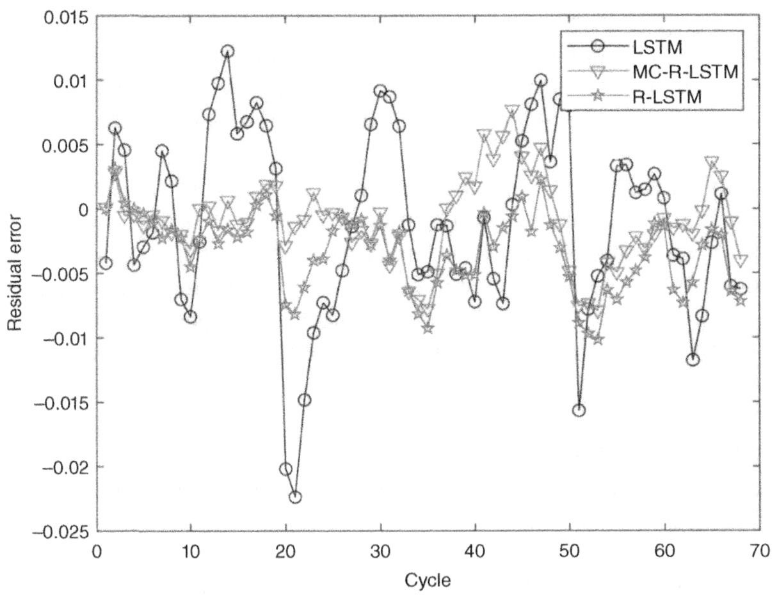

Figure 6.18 Error comparison of the fluctuation component under different models.

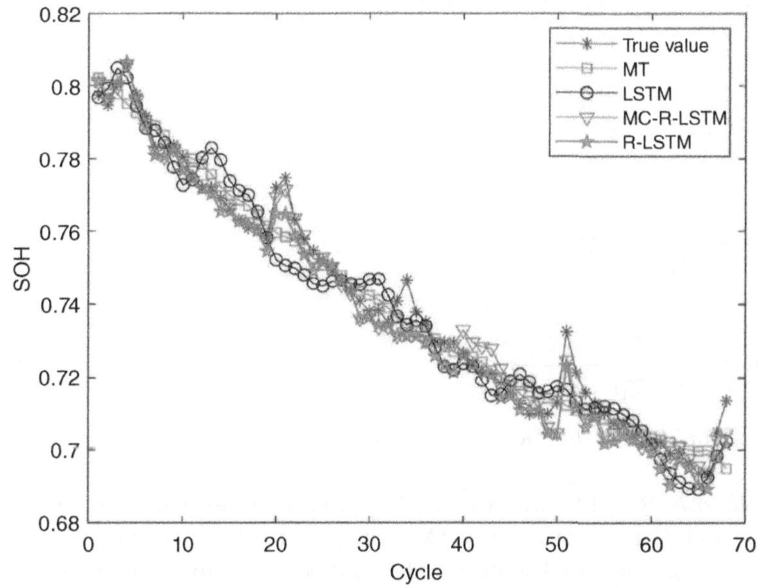

Figure 6.19 Comparison of SOH predictions under different models.

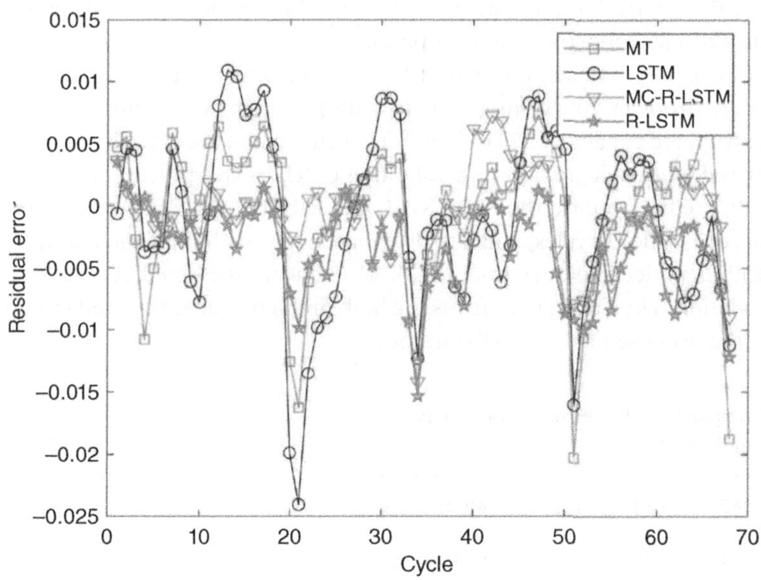

Figure 6.20 SOH prediction error comparison under different models.

The evaluation metrics ME, MAE, MAPE, and RMSE are computed according to Eqs. (6.58)–(6.60) and Eq. (6.56):

$$\mathrm{ME} = \frac{1}{N}\sum_{i=1}^{N}(y_i - \hat{y}_i), \tag{6.58}$$

$$\mathrm{MAE} = \frac{1}{N}\sum_{i=1}^{N}|y_i - \hat{y}_i|, \tag{6.59}$$

$$\mathrm{MAPE} = \frac{100\%}{N}\sum_{i=1}^{N}\left|\frac{y_i - \hat{y}_i}{y_i}\right|. \tag{6.60}$$

The comparison of SOH prediction errors for the four models is provided in Table 6.7.

From Table 6.7, R-LSTM improves SOH prediction accuracy by 36.2%, 35.6%, 34.2%, and 31.1% relative to LSTM in terms of ME, MAE, MAPE, and RMSE, respectively. Compared with the MT method, the improvements reach 14.3%, 13.6%, 11.9%, and 15.0%. Furthermore, MC-R-LSTM achieves an additional 30.0%, 23.7%, 23.1%, and 21.6% improvement over R-LSTM for the same four metrics. These results clearly demonstrate that MC-R-LSTM provides markedly higher accuracy and stronger robustness in SOH prediction.

In the NASA battery dataset, the end of life is defined as the point where the battery capacity drops to 70% of its rated capacity [102]. For the battery considered here with a rated capacity of 2 A · h, failure occurs when the capacity reaches 1.38 A · h. Based on this degradation threshold, the RUL is computed.

For the RUL prediction task, the MT, LSTM, and R-LSTM models produce prediction errors of 1 cycle, 3 cycles, and 1 cycle, respectively. Although each of these methods exhibits noticeable deviation, MC-R-LSTM accurately predicts the RUL with zero deviation. This further confirms the high prediction accuracy and strong reliability of the proposed MC-R-LSTM framework.

Table 6.7 Comparison of SOH prediction errors under different models.

Evaluation	MT	LSTM	R-LSTM	MC-R-LSTM
ME	0.0035	0.0047	0.0030	0.0021
MAE	0.0044	0.0059	0.0038	0.0029
MAPE	0.0059	0.0079	0.0052	0.0040
RMSE	0.0060	0.0074	0.0051	0.0040

6.6 Concluding Remarks

This chapter has explored advanced approaches for the prediction of the RUL of batteries, with a particular emphasis on the integration of filtering techniques and AI. These techniques aim to improve both the precision and robustness of state estimation, while coping with the complex, time-varying and uncertain operating conditions that characterize modern battery systems. The DCR-ZKF was introduced as an effective strategy for state estimation, adept at managing the uncertainties inherent in battery performance. By reducing computational complexity while maintaining high levels of estimation accuracy, the DCR-ZKF offers a practical solution for real-time battery management. The DEE-PF) was proposed as an innovative method for RUL prediction, combining particle filtering with the double exponential model to better capture the degradation processes of lithium-ion batteries. This approach addresses the nonlinear nature of battery degradation more effectively than traditional models, providing more accurate RUL estimates. Furthermore, the integration of anti-aliasing filters with LSTM networks was presented as a hybrid approach to RUL prediction. This combination enables more robust handling of high-frequency fluctuations and noise within battery degradation data, leading to enhanced prediction performance. By mitigating the influence of outliers and smoothing HFCs, this method provides a refined and reliable framework for estimating RUL.

The techniques discussed herein offer valuable contributions to the ongoing refinement of BMSs, particularly in the context of improving the performance and longevity of energy storage systems. These advancements provide both theoretical insights and practical tools that are critical for applications ranging from electric vehicles to large-scale energy storage solutions. As battery technologies evolve, these methods are poised to be of central importance to advancing the capabilities of future BMSs. In conclusion, the integration of filtering methods and AI-driven techniques outlined in this chapter lays a robust foundation for further research and development in battery state estimation. The application of these methods will be key in advancing the efficiency, safety, and longevity of energy storage systems, contributing to the broader goal of sustainable energy solutions.

7

Summary and Future Outlook

7.1 Summary

As the world moves towards carbon neutrality, battery technology plays an increasingly critical role in applications such as new energy vehicles, renewable energy storage, and smart grids. Lithium-ion batteries, known for their high energy density, long cycle life, and efficient rechargeability, have become the cornerstone of energy storage and electric transportation. However, despite the significant advancements in battery technology, several challenges remain in areas such as manufacturing, state estimation, and fault diagnosis. In particular, in the manufacturing process, addressing the issues of non-stationary environments, improving production capacity, forecasting accuracy, and ensuring product consistency have become pressing concerns in the battery industry. Additionally, accurate prediction of battery state of health (SOH) and remaining useful life (RUL) continues to be a challenge, particularly due to the complex degradation processes and environmental variables involved. Traditional methods often assume Gaussian noise and known system models, but battery systems typically exhibit high nonlinearity and complex noise characteristics, making these methods inadequate for real-world applications. Therefore, the integration of advanced filtering techniques and artificial intelligence (AI) strategies has become a key solution to these challenges.

This book combines set-membership filtering techniques and AI approaches to provide innovative solutions to key problems in battery manufacturing and state estimation. Specifically, through set-membership filtering methods, such as zonotope filtering, Kalman filtering, and particle filtering, this book addresses the uncertainties and complex noise environments present in battery systems, enhancing the accuracy and robustness of state estimation for key parameters such as state of charge (SOC), SOH, and RUL. These methods are capable of providing more accurate and reliable state estimations in the face of nonlinear degradation processes, time-varying environmental conditions, and unknown noise, which are typical in battery systems.

In the field of battery manufacturing, this book proposes a data-driven model for optimizing manufacturing capacity predictions. By combining Q-learning and variational mode decomposition (VMD), the key parameters involved in the production process are dynamically adjusted, ensuring high precision in manufacturing capacity forecasting even in non-stationary production environments. The introduction of Q-learning allows the model to adapt to frequently changing environmental conditions in the manufacturing process, optimizing the model's performance in real-time. For example, when combined with VMD, the proposed Q-VMD-LSTM method achieves highly accurate capacity predictions in the presence of production fluctuations. This method not only forecasts capacity effectively based on the fluctuations in the manufacturing process but also provides precise data for production scheduling, resource allocation, and capacity planning. This innovative method is particularly beneficial in adapting to the uncertainties and fluctuations present in battery production, helping manufacturers meet the increasing market demand while ensuring production efficiency and product quality.

In the area of battery state analysis, this book emphasizes the combination of AI techniques and set-membership filtering methods. By utilizing zonotopic filtering, the method is able to perform high-precision SOC and SOH estimation in uncertain and noisy environments. Zonotopic filtering uses geometric sets to represent uncertainties in the system's state, allowing it to handle the nonlinearity and noise fluctuations that are characteristic of battery systems. This method enables real-time capture of the complex behaviors of batteries during charging and discharging processes, particularly in the presence of temperature variations or aging effects. For example, in the health management of electric vehicle (EV) batteries, zonotopic filtering provides real-time monitoring of the battery's charge state and health status, offering reliable data support for battery management systems. Additionally, by incorporating particle swarm optimization (PSO) and clustering optimization techniques, the precision of health status monitoring and fault diagnosis has been further improved, providing powerful tools for fault detection, maintenance, and prediction. Through PSO, the system can adaptively adjust its state estimation process to the dynamically changing environment, ensuring long-term stable monitoring of battery states.

For RUL prediction, this book introduces methods such as the dynamic complexity reduction zonotopic Kalman filter (DCR-ZKF) and double exponential empirical particle filter (DEE-PF). These methods dynamically capture nonlinear features in the battery's degradation process, accurately predicting the RUL of the battery. The DCR-ZKF method achieves a balance between short-term and long-term predictions through multi-scale analysis, effectively tracking the battery's degradation trajectory, thus avoiding overestimations or underestimations of the RUL. Furthermore, the DEE-PF method, using a double-exponential model, accurately predicts both short-term and long-term battery states through a reverse update algorithm, further enhancing the accuracy of RUL predictions. These methods are valuable for making timely maintenance decisions, preventing battery failures during critical usage stages, and extending the battery's lifespan.

The contributions of this book not only enrich the theoretical framework for battery health management and fault diagnosis but also provide innovative solutions for practical applications in manufacturing and operational management. By integrating set-membership filtering techniques with AI strategies, this work drives the intelligent advancement of battery systems in manufacturing, state analysis, and fault diagnosis, serving as a springboard for the further development of smart battery management systems. Looking ahead, with the continuous progress of battery and AI technologies, the integration of intelligent filtering methods will play an increasingly significant role in the production, maintenance, and management of battery systems, promoting the sustainable development of energy storage and electric transportation technologies.

In conclusion, this book provides innovative perspectives and methods for intelligent battery manufacturing and health management. It not only improves the accuracy of performance predictions but also guides the optimization and intelligent development of battery systems. As new technologies continue to emerge, intelligent battery management will become a key trend in the development of battery systems, driving the transformation and upgrading of global energy technologies.

7.2 Future Outlook

Despite the significant progress made in this book regarding state estimation, health management, and fault diagnosis in battery systems, there are still certain limitations. The primary limitation lies in the fact that this research focuses mainly on linear systems and simple nonlinear systems that are linearized for processing. However, in real-world production and applications, many battery systems exhibit highly complex nonlinear behaviors that are not fully addressed by current methods. For example, the electrochemical processes in batteries, material degradation characteristics, and various operating environmental factors all contribute to nonlinear changes in battery behavior. The existing methods often fail to fully capture these complex nonlinearities. Therefore, future research can focus on further advancing nonlinear modeling techniques, particularly by integrating research in battery electrochemistry and material characteristics, to achieve more accurate modeling and state estimation of battery systems. By incorporating more advanced nonlinear dynamic modeling methods, such as deep learning-based nonlinear system modeling, chaos theory, and adaptive nonlinear filtering approaches, it is possible to improve the adaptability and prediction accuracy of existing methods in complex battery systems.

Moreover, while AI methods such as deep learning and reinforcement learning have achieved significant results in battery health monitoring, degradation prediction, and other areas, these methods still rely heavily on large volumes of historical data for training. In real-world applications, especially for new or less common types of batteries and operational environments, there may be insufficient data for model training. Therefore, future research should consider using more flexible data augmentation techniques and transfer learning methods, enabling effective training with smaller datasets and improving the model's generalization ability. For instance, combining virtual simulation data with real experimental data and applying transfer learning to adapt models for different working environments or battery types will help improve the real-time performance and robustness of battery management systems.

In the domain of battery manufacturing, the challenges remain. The battery production process is highly complex and subject to material quality fluctuations, manufacturing process instability, and other factors that lead to inconsistency in battery performance. These issues directly impact battery lifespan and safety. Thus, ensuring consistency throughout the production process, optimizing manufacturing capacity forecasting, and improving quality control are crucial for future research. Real-time data analysis and intelligent optimization techniques will play a key role in overcoming these challenges. By integrating advanced data-driven methods with physical models, the ability to dynamically adjust key parameters during the manufacturing process will improve production efficiency and product quality.

Regarding battery health status estimation, future research could combine more complex electrochemical models with AI-driven data models to further improve the accuracy of SOH and RUL predictions. While current methods are effective in predicting battery health, they are still limited when faced with complex degradation patterns under different battery materials and operational environments. For example, the degradation modes of solid-state batteries differ significantly from those of liquid-state batteries, and current models may fail to capture the degradation behavior of solid-state batteries under specific conditions. Therefore, future research could combine electrochemical models, material degradation models, and deep learning-based multimodal learning methods to enable precise predictions for different types of batteries.

Another key area for future research lies in the environmental adaptability of battery systems during real-world operation. Current battery health management methods are primarily based on theoretical models and data analysis under laboratory conditions, lacking the ability to adapt to changes in temperature, humidity, charge/discharge rates, and other environmental variables. Future research should focus on integrating environmental sensing systems with battery management systems to dynamically monitor the impact of environmental changes on battery performance. For example, by incorporating real-time environmental data, the charging and discharging strategies of batteries can be adjusted dynamically, thereby extending battery lifespan and improving battery safety and reliability under extreme conditions.

In summary, while this book provides innovative insights into intelligent battery manufacturing and health management, there are still many challenges to overcome. Future research should focus on enhancing the integration of nonlinear modeling, electrochemical processes, and material characteristics to achieve more accurate battery system state estimation. By combining advanced filtering techniques, AI, and nonlinear dynamic modeling, future research will enable better optimization of battery systems, leading to more efficient, safer, and sustainable energy storage solutions. As battery technologies and intelligent management systems continue to evolve, future battery systems will become more powerful, safe, and efficient, providing strong support for global energy transitions.

References

1 Olukayode Fatoki, Habeeb Mohammed, Santosh Kumar Parupelli, Alex Mathew, Manpreet Kaur, Amir Rehmat, Sahil Muhammed, Bishnu P. Bastakoti, and Salil Desai. Review of recent advances in lithium-ion batteries: Sources, extraction methods, and industrial uses. *Batteries*, 11(12):433, 2025.

2 Xiang Yun, Xin Zhang, Chao Wang, and Xingming Fan. A review on state of charge estimation methods for lithium-ion batteries based on data-driven and model fusion. *Journal of Energy Storage*, 129:117389, 2025.

3 Kamyar Nikzadfar, Thomas Stone, Jon Caine, Mohammadali Abbasian, Meisam Babaie, Rajib Goswami, Nastaran Ghanami, Muhammad Sheikh, and Mohammad Azadi. *Battery Electric Vehicles: Technologies, Challenges and Future Prospects for Sustainable Transportation*, pages 99–185. Springer Nature, Singapore, 2025.

4 Shuo Cheng. A hybrid deep learning method for the estimation of the state of health of lithium-ion batteries. *International Transactions on Electrical Energy Systems*, 2025:1–10, 2025.

5 Pablo A. García-Salaberri. Chapter 14—Battery manufacturing—from laboratory to industry—challenges. In Amadou Belal Gueye, Hanna J. Maria, Nandakumar Kalarikkal, Modou Fall, Arul Manuel Stephan, and Sabu Thomas (eds.), *Nanostructured Materials Engineering and Characterization for Battery Applications*, pages 465–489. Elsevier, 2024.

6 Guangshu Xia, Chenyu Jia, Yuanhao Shi, Jianfang Jia, Xiaoqiong Pang, Jie Wen, and Jianchao Zeng. Remaining useful life prediction of lithium-ion batteries by considering trend filtering segmentation under fuzzy information granulation. *Energy*, 318:134810, 2025.

7 Xiaodong Xu, Shengjin Tang, Chuanqiang Yu, Jian Xie, Xuebing Han, and Minggao Ouyang. Remaining useful life prediction of lithium-ion batteries based on Wiener process under time-varying temperature condition. *Reliability Engineering & System Safety*, 214:107675, 2021.

8 Kuo Yang, Shunli Wang, Lei Zhou, Carlos Fernandez, and Frede Blaabjerg. A critical review of AI-based battery remaining useful life prediction for energy storage systems. *Batteries*, 11(10):376, 2025.

9 Aditya Tulsyan, Yiting Tsai, R. Bhushan Gopaluni, and Richard D. Braatz. State-of-charge estimation in lithium-ion batteries: A particle filter approach. *Journal of Power Sources*, 331:208–223, 2016.

10 J. Lu, R. Xiong, J. Tian, C. Wang, and F. Sun. Deep learning to estimate lithium-ion battery state of health without additional degradation experiments. *Nature Communications*, 14:2760, 2023.

11 Jingyuan Xue, Xiaozhen Zhao, Dongjing Jiang, Qingchong Jiao, Redouane E. L. Bouchtaoui, and Jianfei Zhang. Survival analysis with machine learning for predicting Li-ion battery remaining useful life, arXiv:2503.13558 [eess.SP], 2025.

12 Sun Jinlei, Liu Wei, Tang Chuanyu, Wang Tianru, Jiang Tao, and Tang Yong. A novel active equalization method for series-connected battery packs based on clustering analysis with genetic algorithm. *IEEE Transactions on Power Electronics*, 36(7):7853–7865, 2021.

13 Kailing Li, Naiming Xie, Nan Chen, Wenjie Dong, and Liudong Gu. Battery degradation prognosis and its health management based on grey iterative modeling. *Engineering Failure Analysis*, 185:110384, 2026.

14 W. Shuai, L. Yiting, Z. Shoubin, C. Lifei, and M. Pecht. Remaining useful life prediction of lithium-ion batteries using a novel particle flow filter framework with grey model. *Scientific Reports*, 15:3311, 2025.

15 Khoa Tran, Bao Huynh, Tri Le, Lam Pham, Vy-Rin Nguyen. HybridoNet-Adapt: A domain-adapted framework for accurate lithium-ion battery RUL prediction. *PLoS One*, 20(10):e0335066.

16 May Htet Htet Khine, Cheong Ghil Kim, and Nattapol Aunsri. A review of Bayesian-filtering-based techniques in rul prediction for lithium-ion batteries. *Journal of Energy Storage*, 111:115371, 2025.

17 X. Yang, H. Zhang, Q. Liu, and G. Jiang. The Li-ion battery industry and its challenges. *Nature Reviews Chemistry*, 9:497–498, 2025.

18 Xiao-Yan Wang, Quan Liu, Qi-Ming Fu, and Le Zhang. Double elite co-evolutionary genetic algorithm. *International Journal of Computer Science and Engineering*, 6(1/2):67–75, 2011.

19 Ziyan Zhao, Shixin Liu, MengChu Zhou, and Abdullah Abusorrah. Dual-objective mixed integer linear program and memetic algorithm for an industrial group scheduling problem. *IEEE/CAA Journal of Automatica Sinica*, 8(6):1199–1209, 2021.

20 Guoqing Chao, Shiliang Sun, and Jinbo Bi. A survey on multiview clustering. *IEEE Transactions on Artificial Intelligence*, 2(2):146–168, 2021.

21 Siamak Talatahari and Mahdi Azizi. Chaos game optimization: A novel metaheuristic algorithm. *Artificial Intelligence Review*, 54(2):917–1004, 2021.

22 Wei Pan, Lichao Ding, Feng Huang, and Y. Sun. Adaptive genetic algorithm based on chaos "micro variation". *Control and Decision*, 36(8): 2042–2048, 2021.

23 Zhan Qi, Maojun Li, hong Mo, Yuhe Xiao, and Fu Liu. Modified genetic algorithm based on state-space model and its convergence analysis. *Control Theory & Applications*, 37(10):2115–2122, 2020.

24 Xuechao Yan, Dongbing Tong, and Qiaoyu Chen. An improved robot path optimization algorithm. *Transducer and Microsystem Technologies*, 39(02):109–112, 2020.

25 International Energy Agency (IEA). Global EV outlook 2024, 2024. Licence: CC BY 4.0.

26 Hasup Song, Injong Gi, Jihyuk Ryu, Yonghwan Kwon, and Jongpil Jeong. Production planning forecasting system based on M5P algorithms and master data in manufacturing processes. *Applied Sciences*, 13(13):7829, 2023.

27 Na Geng and Zhibin Jiang. A review on strategic capacity planning for the semiconductor manufacturing industry. *International Journal of Production Research*, 47(13):3639–3655, 2009.

28 H. Huang, Y. Tian, and Z. F. Tao. Multi-rule combination prediction of compositional data time series based on multivariate fuzzy time series model and its application. *Expert Systems with Applications*, 238:121966, 2024.

29 G. Dudek. STD: A seasonal-trend-dispersion decomposition of time series. *IEEE Transactions on Knowledge and Data Engineering*, 35(10):10339–10350, 2023.

30 A. Wang, P. Qin, X. M. Sun, and Y. S. Li. An automatic parameter setting variational mode decomposition method for vibration signals. *IEEE Transactions on Industrial Informatics*, 20(2):2053–2062, 2024.

31 Y. Yin, L. Feng, Q. Zhang, Y. Wang, L. Song, H. Chi, and Z. H. Wen. A wear particle detection method based on coaxial array electrostatic sensor and VMD-SR-DTW model. *IEEE Transactions on Instrumentation and Measurement*, 73:9514611, 2024.

32 Krishna Prakash Natarajan and Jai Govind Singh. Day-ahead residential load forecasting based on variational mode decomposition and hybrid deep networks with Granger causality feature selection. In *2024 IEEE 3rd Industrial Electronics Society Annual On-Line Conference (ONCON)*, pages 1–6, 2024.

33 Zhuoqun Zou, Jing Wang, E. Ning, Can Zhang, Zhaocai Wang, and Enyu Jiang. Short-term power load forecasting: An integrated approach utilizing variational mode decomposition and TCN–BiGRU. *Energies*, 16(18):6625, 2023.

34 A. A. M. Ahmed, N. Bailek, L. Abualigah, K. Bouchouicha, A. Kuriqi, A. Sharifi, P. Sareh, P. Mishra, I. Colak, and E. S. M. El-Kenawy. Global control of electrical supply: A variational mode decomposition-aided deep learning model for energy consumption prediction. *Energy Reports*, 10:2152–2165, 2023.

35 T. Muto, T. Katayama, Y. Kita, H. Yamaba, K. Aburada, and N. Okazaki. Expansion of application scope and addition of a function for operations into

BWDM which is an automatic test cases generation tool for VDM++ specification. *Journal of Robotics Networking and Artificial Life*, 9(3): 255–262, 2022.

36 Z. Bian, J. Zhong, J. Dominic, G. E. Christensen, C. R. Hatt, and N. S. Burris. Validation of a robust method for quantification of three-dimensional growth of the thoracic aorta using deformable image registration. *Medical Physics*, 49(4):2514–2530, 2022.

37 M. Singh. A neural network-guided correlation thresholding technique for wireless sensor networks. *Measurement*, 239:115408, 2025.

38 A. A. Kroshchanka, V. A. Golovko, and M. Chodyka. Method for reducing neural-network models of computer vision. *Pattern Recognition and Image Analysis*, 32(2):294–300, 2022.

39 C. Xie, Z. Li, and R. Song. Deep-LSTM-based human robot collaboration control with static-dynamic calibration. *IEEE Transactions on Industrial Electronics*, 71(12):16165–16175, 2024.

40 K. Huang, K. Wei, F. Li, C. Yang, and W. H. Gui. LSTM–MPC: A deep learning based predictive control method for multimode process control. *IEEE Transactions on Industrial Electronics*, 70(11):11544–11554, 2024.

41 R. Dornberger, P. Chintala, and T. Hanne. Robotic path planning by Q learning and a performance comparison with classical path finding algorithms. *International Journal of Mechanical Engineering and Robotics Research*, 11(6):373–378, 2022.

42 X. Qi and B. Xu. Hyperparameter optimization of neural networks based on Q-learning. *SIViP*, 17:1669–1676, 2023.

43 Philipp Andelfinger, Jieyi Bi, Qiuyu Zhu, Jianan Zhou, Bo Zhang, Fei Fei Zhang, Chew Wye Chan, Boon Ping Gan, Wentong Cai, and Jie Zhang. Learning to optimize capacity planning in semiconductor manufacturing. In *Methods and Applications for Modeling and Simulation of Complex Systems. AsiaSim 2025. Communications in Computer and Information Science*, pages 67–79, 2025.

44 R. S. Sutton and A. G. Barto. *Reinforcement learning: An introduction*. The MIT Press, 2nd edition, 2018.

45 T. Alamo, J. M. Bravo, and E. F. Camacho. Guaranteed state estimation by zonotopes. In *42nd IEEE International Conference on Decision and Control*, pages 5831–5836, 2003.

46 V. T. H. Le, C. Stoica, T. Alamo, E. F. Camacho, and D. Dumur. Zonotope-based set-membership estimation for multi-output uncertain systems. In *2013 IEEE International Symposium on Intelligent Control*, pages 212–217, 2013.

47 Y. Wang, Z. Wang, V. Puig, and G. Cembrano. Zonotopic set-membership state estimation for discrete-time descriptor LPV systems. *IEEE Transactions on Automatic Control*, 64(5):2092–2099, 2019.

48 S. Ben Chabane, C. Stoica Maniu, T. Alamo, E. F. Camacho, and D. Dumur. Improved set-membership estimation approach based on zonotopes and ellipsoids. In *2014 European Control Conference*, pages 993–998, 2014.

49 J. M. Cabello, X. Roboam, S. Junco, E. Bru, and F. Lacressonniere. Scaling electrochemical battery models for time-accelerated and size-scaled experiments on test-benches. *IEEE Transactions on Power Systems*, 32(6):4233–4240, 2017.

50 Brenner S. Rego, Diego Locatelli, Davide M. Raimondo, and Guilherme V. Raffo. Joint state and parameter estimation based on constrained zonotopes. *Automatica*, 142:110425, 2022.

51 R. Xiong, R. Yang, Z. Chen, W. Shen, and F. Sun. Online fault diagnosis of external short circuit for lithium-ion battery pack. *IEEE Transactions on Industrial Electronics*, 67(2):1081–1091, 2020.

52 O. A. Stepanov, V. A. Vasiliev, M. V. Basin, V. A. Tupysev, and Y. A. Litvinenko. Efficiency analysis of polynomial filtering algorithms in navigation data processing for a class of nonlinear discrete dynamical systems. *IET Control Theory & Applications*, 15(2):248–259, 2021.

53 Christophe Combastel. Zonotopes and Kalman observers: Gain optimality under distinct uncertainty paradigms and robust convergence. *Automatica*, 55:265–273, 2015.

54 Ramon Moore and Weldon Lodwick. Interval analysis and fuzzy set theory. *Fuzzy Sets and Systems*, 135(1):5–9, 2003.

55 C. Combastel. An extended zonotopic and Gaussian Kalman filter merging set-membership and stochastic paradigms: Toward non-linear filtering and fault detection. *Annual Reviews in Control*, 42:232–243, 2016.

56 Esmaeil Siahlooei and Seyed Abolfazl Shahzadeh Fazeli. An application of interval arithmetic for solving fully fuzzy linear systems with trapezoidal fuzzy numbers. *Advances in Fuzzy Systems*, 2018:2104343, 2018.

57 Ziyun Wang, Guixiang Xu, Zixing Liu, Yan Wang, and Zhicheng Ji. Orthotopic-filtering-based fault diagnosis algorithms for nonlinear systems with slowly varying faults. *Journal of The Franklin Institute*, 357(9):5610–5639, 2020.

58 J. W. Wei, G. Z. Dong, and Z. H. Chen. Lyapunov-based thermal fault diagnosis of cylindrical lithium-ion batteries. *IEEE Transactions on Industrial Electronics*, 67(6):4670–4679, 2020.

59 S. F. Wang, Y. Luo, L. Dong, and L. G. Liang. Research on offline identification of equivalent model parameters of second order thevenin lithium batteries. *Electronic Design Engineering*, 26(9):46–49, 2018.

60 R. Cheng and J. Huang. Center difference set membership filter by zonotopes for nonlinear system. In *International Conference on Modelling, Identification and Control*, pages 1–6, 2018.

61 Jun Hu, Zidong Wang, and Guo-Ping Liu. Delay compensation-based state estimation for time-varying complex networks with incomplete observations and dynamical bias. *IEEE Transactions on Cybernetics*, 52(11):12071–12083, Nov 2022.

62 Jie Kong, Ben Niu, and Zhi Wang. Adaptive two-stage unscented Kalman filter for state of charge estimation of lithium-ion batteries. *IEEE Transactions on Systems, Man, and Cybernetics: Systems*, 51(11):6721–6732, 2021.

63 T. Alamo, J. M. Bravo, M. J. Redondo, and E. F. Camacho. A set-membership state estimation algorithm based on DC programming. *Automatica*, 44(1):216–224, Jan 2008.

64 Jos Elfring, Elena Torta, and René van de Molengraft. Particle filters: A hands-on tutorial. *Sensors*, 21(2):438, 2021.

65 Amirhossein Taghvaei and Prashant G. Mehta. A survey of feedback particle filter and related controlled interacting particle systems (CIPS). *Annual Reviews in Control*, 55:356–378, 2023.

66 Ahmed G. Gad. Particle swarm optimization algorithm and its applications: A systematic review. *Archives of Computational Methods in Engineering*, 29:2531–2561, 2022.

67 Z. Y. Wang, L. Cheng, Y. Wang, and Zhi-cheng Ji. Orthometric hyperparallel spatial directional expansion filtering based fault diagnosis method [in Chinese]. *Control and Decision*, 37(12):3223–3232, 2022.

68 C. E. Valero and R. Paulen. Effective recursive set membership state estimation for robust linear MPC. *IFAC-PapersOnLine*, 52(1):486–491, 2019.

69 Andre Bigand and Olivier Colot. Membership function construction for interval-valued fuzzy sets with application to Gaussian noise reduction. *Fuzzy Sets and Systems*, 286:66–85, Mar 2016.

70 Zixing Liu, Ziyun Wang, Yan Wang, and Zhicheng Ji. Sensor fault estimation based on the constrained zonotopic Kalman filter. *International Journal of Robust and Nonlinear Control*, 31(12):5984–6006, Aug 2021.

71 Matthias Althoff, Olaf Stursberg, and Martin Buss. Computing reachable sets of hybrid systems using a combination of zonotopes and polytopes. *Nonlinear Analysis: Hybrid Systems*, 4(2):233–249, Jan 2010.

72 L. Chisci, A. Garulli, and G. Zappa. Recursive state bounding by parallelotopes. *Automatica*, 32(7):1049–1055, Jul 1996.

73 Marco Casini, Andrea Garulli, and Antonio Vicino. Set membership state estimation for discrete-time linear systems with binary sensor measurements. *Automatica*, 159:111396, 2024.

74 Yuchen Zhang, Bo Chen, and Li Yu. Distributed zonotopic estimation for interconnected systems: A fusing overlapping states strategy. *Automatica*, 155:111144, 2023.

75 Wentao Tang, Zhenhua Wang, and Yi Shen. Fault detection and isolation for discrete-time descriptor systems based on h_-/l_∞ observer and zonotopic residual evaluation. *International Journal of Control*, 93(8):1867–1878, 2020.

76 Tareq M. Shami, Ayman A. El-Saleh, Mohammed Alswaitti, Qasem Al-Tashi, Mhd Amen Summakieh, and Seyedali Mirjalili. Particle swarm optimization: A comprehensive survey. *IEEE Access*, 10:10031–10061, 2022.

77 W. Zhang, Z. Wang, S. Guo, and Y. Wang. Interval estimation of sensor fault based on zonotopic Kalman filter. *International Journal of Control*, 94(6):1641–1650, 2021.

78 Z. X. Liu, Z. Y. Wang, Y. Wang, and Z. C. Ji. Optimal zonotopic Kalman filter-based state estimation and fault-diagnosis algorithm for linear discrete-time system with time delay. *International Journal of Control, Automation and Systems*, 20(6):1757–1771, 2022.

79 Y. Wang, Z. Wang, V. Puig, and G. Cembrano. Zonotopic fault estimation filter design for discrete-time descriptor systems. *IFAC-PapersOnLine*, 50(1): 5055–5060, 2017.

80 P. Wang, B. Xue, J. J. Liang, and M. Zhang. Differential evolution-based feature selection: A niching-based multi-objective approach. *IEEE Transactions on Evolutionary Computation*, 27(2):296–310, April 2023.

81 S. Lin, Y. L. Zhang, Z. H. Ding, and W. Q. Niu. State estimation of automated straddle carriers via improved set-membership filtering approach [in Chinese]. *Control and Decision*, 39(1):129–136, 2024.

82 Huan Xu, Feng Ding, and Benoit Champagne. Joint parameter and time-delay estimation for a class of nonlinear time-series models. *IEEE Signal Processing Letters*, 29:947–951, 2022.

83 M. Althoff and B. H. Krogh. Avoiding geometric intersection operations in reachability analysis of hybrid systems. *IEEE Transactions on Automatic Control*, 59(2):508–522, 2014.

84 Zhenhua Wang, Yilian Zhang, Mouquan Shen, and Yi Shen. Ellipsoidal set-membership filtering for discrete-time linear time-varying systems. *IEEE Transactions on Automatic Control*, 68(9):5767–5774, 2023.

85 Zhongyang Fei, Liu Yang, Xi-Ming Sun, and Shunqing Ren. Zonotopic set-membership state estimation for switched systems with restricted switching. *IEEE Transactions on Automatic Control*, 67(11):6127–6134, 2022.

86 C. Combastel. A state bounding observer for uncertain nonlinear continuous-time systems based on zonotopes. *IEEE Transactions on Automatic Control*, 48(11):2074–2078, 2003.

87 Teodoro Alamo, Jose M. Bravo, and Eduardo F. Camacho. Guaranteed state estimation by zonotopes. *Automatica*, 41(6):1035–1043, 2005.

88 Y. Yang and O. Stursberg. Set-membership state estimation for nonlinear systems with bounded uncertainties using zonotopes. *Journal of Process Control*, 22(2):314–325, 2012.

89 B. Saha, K. Goebel, and J. Christophersen. Comparison of prognostic algorithms for estimating remaining useful life of batteries. *Transactions of the Institute of Measurement and Control*, 31(3–4):293–308, 2009.

90 Q. Zhang and R. E. White. Capacity fade analysis of a lithium-ion cell. *Journal of Power Sources*, 179(2):793–798, 2008.

91 W. He, N. Williard, and M. Osterman. Remaining useful performance analysis of batteries. In *Proceedings of the 2011 IEEE Conference on Prognostics and Health Management*, pages 1–6, 2011.

92 D. Liu, J. Zhou, and L. Guo. Survey on lithium-ion battery health assessment and cycle life estimation. *Chinese Journal of Scientific Instrument*, 36(1):1–16, 2015. In Chinese.

93 B. E. Olivares, M. A. Cerda Munoz, M. E. Orchard, and J. F. Silva. Particle-filtering-based prognosis framework for energy storage devices with a statistical characterization of state-of-health regeneration phenomena. *IEEE Transactions on Instrumentation and Measurement*, 62(2):364–376, 2013.

94 F. Wang, M. Lu, Q. Zhao, and Z. Yuan. Particle filter algorithm. *Chinese Journal of Computers*, 37(8):1679–1694, 2014. In Chinese.

95 K. Huang, H. Ding, Y. F. Guo, and H. J. Tian. Life prediction of lithium-ion battery based on data preprocessing and long and short term memory neural network. *Transactions of China Electrotechnical Society*, 37(15):3753–3766, 2022. In Chinese.

96 H. F. Liang, P. Yuan, and Y. J. Gao. Remaining useful life prediction of lithium-ion battery based on CNN-Bi-LSTM network. *Electric Power Automation Equipment*, 41(10):213–219, 2021. In Chinese.

97 C. L. Zhang, Y. G. He, and L. F. Yuan. Prediction approach for remaining useful life of lithium-ion battery based on EEMD and MKRVM. *Proceedings of the Chinese Society of Universities for Electric Power System and Automation*, 30(7):38–44, 2018. In Chinese.

98 L. Z. Yi, Z. G. Zhang, C. D. Fan, X. G. Luo, W. Li, and W. H. Liu. Life prediction of lithium battery based on EEMD-GSGRU. *Energy Storage Science and Technology*, 9(5):1566–1573, 2020. In Chinese.

99 Thomas Chevet, Thach Ngoc Dinh, Julien Marzat, Zhenhua Wang, and Tarek Raïssi. Zonotopic Kalman filter-based interval estimation for discrete-time linear systems with unknown inputs. *IEEE Control Systems Letters*, 6:806–811, 2022.

100 X. Su, S. Wang, and M. Pecht. Interacting multiple model particle filter for prognostics of lithium-ion batteries. *Microelectronics Reliability*, 70:59–69, 2017.

101 W. Chen, Y. Cai, and Y. Su. Remaining useful life prediction of lithium-ion battery based on improved particle filtering. *China Measurement & Test*, 47(7):148–153, 2021.

102 T. Z. Hu and J. B. Yu. Life prediction of lithium-ion batteries based on multiscale decomposition and deep learning. *Journal of Zhejiang University (Engineering Science)*, 53(10):1852–1864, 2019.

103 D. Wu, J. B. Huang, and M. R. Zhong. Prediction of metal futures price volatility and empirical analysis based on symbolic time series of high-frequency. *Transactions of Nonferrous Metals Society of China*, 30(6):1707–1716, 2020.

Index

Note: Page numbers in *italics* and **bold** refers to figures and tables respectively.

u

v

z

Wiley The manufacturer's authorized representative according to the EU General Product Safety Regulation is Wiley-VCH GmbH, Boschstr. 12, 69469 Weinheim, Germany, e-mail: Product_Safety@wiley.com.

Printed and bound by CPI Group (UK) Ltd, Croydon, CR0 4YY
15/09/2026
02217457-0002